WHAT SCIENCE IS
AND HOW IT WORKS

WHAT SCIENCE IS
AND HOW IT WORKS

GREGORY N. DERRY

PRINCETON UNIVERSITY PRESS

PRINCETON, NEW JERSEY

Library of Congress Cataloging-in-Publication Data

Derry, Gregory Neil, 1952–
What science is and how it works /
Gregory N. Derry.
p.   cm.
Includes bibliographical references and index.
ISBN 0-691-05877-6 (cloth : alk. paper)
1. Science.   I. Title.
Q158.5.D47   1999   500—dc21   99-17186   CIP

To Paula and Rebecca

CONTENTS

PREFACE

SCIENCE, like many other topics, is much more interesting if it makes sense to you. I wrote this book because science is extraordinarily interesting to me, and I want to share that interest with other people. My goal for the book is to convey the foundations of my own understanding of science, which I have acquired over an extended period of time. Scholars argue over whether science is a body of knowledge, a collection of techniques, a social and intellectual process, a way of knowing, a strictly defined method, and so forth. These arguments are not very interesting to me, since I accept all of these elements as valid partial visions of science. In one guise or another, they all appear somewhere in the book. My other motivation for writing the book is to show that science, as well as being interesting, is also important. A significant part of our culture, our economy, and our environment are entangled with science in profound ways. To comprehend the world we live in without some grasp of science is difficult. Crucial issues are at stake, and these issues require an understanding of science in order to approach them intelligently.

The audience for this book is anybody with some curiosity about the issues I explore. No particular background is assumed. In writing, I especially had in mind a reader who enjoys ideas but hasn't studied the sciences in any depth. People who have a scientific background will also find the book of interest, but I primarily had in mind people who are not experts. In fact, my underlying assumption is that you don't need any particular expertise to have a genuine understanding of what science is and how science works.

In order to keep the scope of the book manageable, I am using the word "science" to mean natural science. (This is merely a convenient convention, not intended to reflect any opinion about the relative worth of the disciplines I'm not including.) The social sciences, mathematics, and engineering are sometimes discussed briefly, but the main focus of the book is on chemistry, biology, physics, and the earth sciences. I have tried to avoid any prejudice in favor of a particular discipline. I have also tried to avoid favoring either the laboratory sciences or the historical/observational sciences. My own background is in physics, and that may have colored my treatment and choice of topics. Nevertheless, I have tried to maintain a broad transdisciplinary flavor.

A number of books already try to explain science to the general public. I would like to articulate why I have written another one and why what I have tried to accomplish is different. My overarching goal is to give the reader *more than just a description* of how other people (scientists) think about the world; I want to communicate this thought process to readers in a way that enables them to *actually engage in a similar thought process.* My other claim to novelty is the distinctive combination of different approaches I've employed: historical narratives, integrative cross-disciplinary ideas and concepts, comparisons with other (nonscientific) endeavors, and characteristically scientific tactics for thinking about the world. Lastly, I have put a lot of effort into presenting substantial ideas in a way that does not oversimplify these ideas into fluff, but also does not bore the reader to death. Of course, I don't want to promise too much. I have covered a lot of ground in just a few hundred pages. For every topic I discuss, multiple volumes have been written. I can only scratch the surface here and try to illuminate the major points of each issue with broad brushstrokes. But despite these limitations, my intention is to get to the heart of the matter in every case.

I have generally avoided expressing personal opinions on controversial issues (social, political, or scientific), opting instead to present all sides as fairly as I could. On the other hand, there are also sections of the book where I have presented views that reflect a broad consensus among many reasonable people, though other opinions may exist. In a few places, I express personal opinions because I could not see any way to avoid it; I have clearly indicated those passages that present no one's thinking but my own.

Finally, because this book contains so many interrelated ideas, I have employed quite a few cross-references throughout. This practice allows readers (optionally) to find useful information and background when unfamiliar ideas appear. My intention is to allow the book to be read in an order other than from beginning to end. If you are one of those readers who is well-adapted to the new electronic age, you can think of these cross-references as hypertext links and pretend you are clicking a mouse as you turn to the indicated page.

. . . . .

I have many debts to acknowledge regarding the creation of this book, which is based on many years of prior work. My thinking during all that time has been influenced by many teachers, colleagues, and friends. Among my teachers, Prof. C. D. Swartz stands out as the first person who introduced me to real science. The rest of my teachers and colleagues who have contributed to my thinking over the years are too numerous to mention. Many of my friends have influenced my thinking in important ways; Scott Wittet, Paul Ferguson, and Christine High deserve special mention.

A number of people have also contributed more directly to the development of the book. Betsy Reeder, Dan Perrine, Fr. Frank Haig, S.J., and Peter French have all read portions of the manuscript, offering both criticism and encouragement. Randy Jones and Helene Perry read through an entire early version, offering me a variety of suggestions for improvement. Two anonymous reviewers carefully and thoroughly read this early version and also a completed later version, in both cases providing me with many corrections and recommendations. Judy Dobler made a very careful and critical reading of some early chapters, and supplied me with a remarkably voluminous set of notes and stylistic comments; everything I wrote afterward was influenced by these suggestions. And Trevor Lipscombe was able to see possibilities for this book that I had not been able to see myself.

I would also like to acknowledge my institution, Loyola College, for providing a sabbatical leave during which the writing of this book was started. There is no possibility that the book could exist if I had not had that unencumbered period of time to focus on it.

Finally, I owe several debts to my family, Paula and Rebecca Derry. They have supported this arduous venture in many ways, including encouragement. My daughter, Rebecca, has checked passages for clarity, caught some typographical and grammatical mistakes, and contributed to the figures. My wife, Paula, has read a great deal of the manuscript, offering incisive critical comments on both content and style. She has also greatly influenced my thinking about a number of key issues for many years prior to the writing of the book. (I should also mention the cats, Katie and Smokey, who amused me by walking across the keyboard as I tried to think of something suitable to write.)

Although the many suggestions I received have improved the book greatly, I must take responsibility for the final product. I have not made all of the changes that have been suggested. In the end, I had to decide what should be included or not. Writing this book has been a lot of hard work, but it has also been very enjoyable. I hope your experience of reading the book is rewarding and congenial.

Baltimore
July 1998

WHAT SCIENCE IS
AND HOW IT WORKS

Prologue

WHAT IS SCIENCE?

> Science is the last step in man's mental development and it
> may be regarded as the highest and most characteristic
> attainment of human culture.
> *(Ernst Cassirer)*

> The belief that science developed solely out of a pursuit of
> knowledge for its own sake is at best only a half truth, and
> at worst, mere self-flattery or self-deception on the part
> of the scientists.
> *(Lewis Mumford)*

AS THE OPENING QUOTATIONS by two noted philosophers indicate, opinions about science span a wide range. But it's not clear whether these two eminent thinkers are really talking about the same thing when they refer to "science." Cassirer is discussing science as an abstract method to bring constancy and regularity to the world. Mumford, in contrast, is considering science as a driver of technology, a method to bring about practical changes in life. Both of these viewpoints contain an element of truth; neither is comprehensive. A simple, brief, and comprehensive way to define science is in fact not so easy to come up with. A colleague of mine recently remarked that the defining characteristic of science is that statements in science must be tested against the behavior of the outside world. This statement is fine as far as it goes, but represents a rather impoverished picture of science. Where are imagination, logic, creativity, judgment, metaphor, and instrumentation in this viewpoint? All these things are a part of what science is.

Science is sometimes taken to be the sum total of all the facts, definitions, theories, techniques, and relationships found in all of the individual scientific disciplines. In other words, science is what is taught in science textbooks. Many beginning science students have this idea. But an opposing opinion, which is becoming increasingly influential, has been expressed in academic circles. In this view, the heart of science is in its methods of investigation and ways of thinking, not in specific facts and results. The science taught in textbooks is a lifeless husk, whereas real science is the activity going on in the laboratories and fieldwork. Once again, both of these ideas have merit while neither can claim to be complete. Methodology without content is at best merely a faint image of science (at worst, it's totally meaningless). And yet the content itself, divorced from the thought processes that create such knowledge, surely can't be all there is to science. After all, this body of scientific results changes from year to year, and may sometimes be unrecognizable from one generation to another. The results of science are inseparably intertwined with its thought processes; both together are needed to understand what science is.

There are many other such debates and contrasting perspectives among scientists and philosophers concerning the true nature of science, and we'll consider a number of them as we go along. For now, though, let's take a rest from these abstractions and look at a small example of science in action. Our example concerns something of interest to almost everyone: food.

Example: Why should you whip a meringue in a copper bowl?

As anyone who has made a lemon meringue pie knows, whipping egg whites results in a somewhat stiff foam (the meringue). A tradition in cooking, which can be traced at least back to the eighteenth century, is that egg whites are best whipped in a copper bowl when making meringues. The meringue turns out creamier and less prone to overbeating if the bowl is made of copper (the creamy meringue also has a somewhat yellowish color). Less elite cooks, like myself, achieve a somewhat similar result by using cream of tartar in the meringue instead of beating it in a copper bowl. The interesting question then presents itself: How and why does using a copper bowl affect the meringue?

To understand the influence of the copper bowl, we must first understand why a meringue forms at all. Why do egg whites make a stiff foam when they are whipped? The answer to this question is related to the composition of the egg white (also called albumen), which is a complex substance containing many different proteins (ovalbumen, conalbumen, ovomucin, lysozyme, etc.) suspended in water. These proteins contain long chains of amino acids twisted together into a compact form. The compact protein structure is maintained by chemical bonds between various parts of the twisted chains, acting as a kind of glue. As you whip the egg whites, these bonds weaken and the amino acid chains start to unfold, mostly due to contact with the air contained within the bubbles you create by whipping. The unfolded chains of *different* protein molecules can then start bonding to each other, eventually forming a latticework of overlapping chains that surrounds the bubble wall. The water in the egg white is also held together within this network of protein chains. The protein network reinforces the bubble walls and so maintains the structural integrity of the foam. And we have a meringue.

If you overbeat the egg whites, however, the meringue turns into curdled lumps floating in a watery liquid. The reason this happens is that the network of protein chains becomes too tightly woven and can no longer hold enough water within its structure. The bonding between chains has become too effective, leaving few bonding sites for water molecules. The protein turns into clumps while the water drains out. Adding a little cream of tartar helps to avoid this unfortunate outcome. The cream of tartar is slightly acidic, contributing excess hydrogen ions that interfere with the cross-chain bonding. With weaker bonding, the meringue is less likely to become overbeaten.

This brings us back to the copper bowls, which confer the same virtue: less likelihood of overbeating. Basing our reasoning on the cream of tartar example, we might guess that the copper bowl somehow increases the acidity of the egg white. But such an increase would be difficult to understand, and in any event a simple measurement of the acidity proves that this idea is wrong. Instead, the answer turns out to be related to the ability of conalbumen, one of the proteins making up egg white, to bind metal ions (in this case, copper) to itself. The copper ions that are incorporated into the conalbumen molecule have a striking effect; they stabilize the coiled structure of the protein, acting to prevent the chains from unfolding. Standard laboratory chemistry experiments had demonstrated this fact many decades ago. Since the conalbumen (with copper added) isn't unfolded, its chains don't take part in the formation of a stable foam. If we assume that a small but significant number of copper atoms are scraped from the sides of the bowl into the egg white, then we have a good possible explanation of why copper bowls help to prevent overbeating.

We can test our explanation. These conalbumen/copper complexes absorb light of certain specific colors. Looking at the light absorbed by meringues, we can find out if they really do have such conalbumen/copper complexes. This test has actually been performed, and light absorption experiments using meringues beaten in a copper bowl do indeed reveal the presence of stable conalbumen/copper molecules. Incidentally, the light absorption properties of the complex give it a characteristic yellow color, and so we also have an explanation for the yellowish color of the meringue. This modest example is far removed from the grand philosophical debates about science, but it nicely illustrates a number of important themes: science is about real things that happen in the world; science tries to provide a coherent understanding of these things; our specific observations must be placed in a more general framework to be understood; interpretations are often based on pictorial models; we often use instruments and measurements to augment our observations; a genuinely coherent picture often leads to predictions of new observations, which serve as tests of how correct our present interpretation is. Most of these themes, as well as many others, will recur throughout the book.

AN OVERVIEW

The first part of the book is about scientific discoveries. More particularly, we examine the question of how discoveries are made. I'm not interested in undertaking a systematic and exhaustive investigation of the sources of scientific discovery, however, and I'm certainly not trying to devise a theory to explain the process of discovery. My firm belief is that there are many, many factors involved in this process, and they vary greatly from one situation to another. My only goal is to illustrate some of these factors by looking at examples. Since I'm looking at particular examples of discoveries, this part of the book is primarily historical. The historical approach allows us to look at the rich context of each discovery, without distorting the narrative to fit into a preconceived notion. On the other hand, I am trying to use each example to illustrate some particular element that played a dominant role in the discovery under discussion (even when several other factors were also important). Some of these dominant elements include: the apprehension of patterns in data; increased power of instrumentation; luck (serendipity); the role of discrepancies; thematic imagination; the hypothetico-deductive method; the consequences of a priori postulates; and inspired flashes of intuition.

In the second part of the book, we shift gears and approach science from quite a different angle. For some time now, it has seemed to me that scientists often approach the world with a rather distinctive kind of thinking process. I don't mean by this that any particular method is applied; rather, I'm referring to a style of looking at questions and approaching problems. Let me illustrate this vague statement with an example. When I was on a jury deciding an automobile accident lawsuit, I was the only person who asked: "What plausible model can we construct for the accident that is consistent with the photographs of the damage?" The other jurors weren't entirely sure what I meant by this. Constructing models is a very typical way for a scientist to think about a situation. Science is often done this way, and scientists naturally extend the practice to other situations. As I said, this practice (thinking in terms of models) is only one example of the style I'm talking about. Another customary approach is to employ quantitative thinking about a situation (for example, "how precisely do I know this number?" or "does the order of magnitude of that number make sense?"). Yet another example is the habit of looking for general principles of known validity against which to judge particular claims. These sorts of characteristic scientific thought processes and approaches are the subject of the second part of the book.

The third part of the book is an endeavor to place science within a broader matrix of ideas. An important part of this undertaking is to look at what science *is* by looking more closely at what science is *not*. Of course, a great deal of human experience and thought lies outside science, but we're mostly concerned with those areas that do have some overlapping interests. For this reason, vast subjects like religion, politics, and ethics are discussed somewhat narrowly, primarily in terms of how they relate to science. On a much different note, we also contrast science with pseudoscience, which might be described as a burlesque of real science (but unfortunately is often taken seriously). Moving from there into controversial territory, we look at some areas where arguments are still raging over whether the topics in question are science or not. Then, after a rather condensed summary of the main ideas and issues in the philosophy of science, we again enter into an intellectual minefield and briefly discuss the arguments of the postmodern critics of science.

In the fourth and final part of the book, we consider some of the broad concepts and ideas important in the sciences. Although each of the individual scientific disciplines has its own central principles (for example, natural selection in biology or plate tectonics in geology), the concepts emphasized in this part of the book are transdisciplinary. In other words, the subjects discussed here cut across disciplinary boundaries and are important in a variety of different sciences. In this way, I hope to show some of the underlying unity of the sciences, which can become lost in the fragmentary treatment of particular results. A prime example of such broadly important concepts is symmetry. Though symmetry is in many ways a mathematical concept, it is significant in art and aesthetics as well as in virtually every science. Another good example is the dependence of volume and surface area on the characteristic size of an object; this too turns out to be important in many areas of science (as well as in practical affairs). Very often in the sciences, a prominent consideration is how something changes. Two of the most common and useful kinds of change are discussed here: linear variation (one thing proportional to another) and exponential variation (growth rate proportional to amount). Profound issues at the heart of many sciences turn on the concepts of order and disorder, which are treated here in some detail. We then round out this part of the book with a discussion of feedback loops and homeostasis in the sciences. The book ends with a brief epilogue in which we will reconsider the question: what is science?

FOR FURTHER READING

Ideas of Science, by Bernard Dixon and Geoffrey Holister, Basil Blackwell, 1984.
On Food and Cooking: The Science and Lore of the Kitchen, by Harold McGee, Macmillan, 1984.
The Game of Science, by Garvin McCain and Erwin M. Segal, Brooks/Cole, 1988.
The Scientific Companion, by Cesare Emiliani, John Wiley & Sons, 1988.
The Scientific Attitude, by Frederick Grinnell, Guilford Press, 1992.
Science and Its Ways of Knowing, edited by John Hatton and Paul B. Plouffe, Prentice-Hall, 1997.

EPIGRAPH REFERENCES: Ernst Cassirer, *An Essay On Man*, Yale University Press, 1962, p. 207. Lewis Mumford, *The Pentagon of Power*, Harcourt Brace Jovanovich, 1970, p. 106.

PART I

EXPLORING THE FRONTIERS OF SCIENCE:

HOW NEW DISCOVERIES ARE MADE

IN THE SCIENCES

Chapter 1

A BIRD'S EYE VIEW: THE MANY ROUTES
TO SCIENTIFIC DISCOVERY

> Now, I am not suggesting that it is impossible to find natural
> laws; but only that this is not done, and cannot be done, by
> applying some explicitly known operation. . . .
> *(Michael Polanyi)*

HOW DOES A SCIENTIST go about making a discovery? The idea that there's a single answer to this question (the "scientific method") persists in some quarters. But many thoughtful people, scientists and science critics alike, would now agree that science is too wide-ranging, multifaceted, and far too interesting for any single answer to suffice. No simple methodology of discovery is available for looking up in a recipe book. To illustrate some of the rich variety in the ways scientists have discovered new knowledge, I have chosen five cases to recount in this chapter: the accidental discovery of x-rays; the flash of intuition leading to the structure of benzene; the calculations through which band structure in solids was discovered; the voyages of exploration inspiring the invention of biogeography; and the observations and experiments resulting in smallpox vaccine.

§1. SERENDIPITY AND METHODICAL WORK:
ROENTGEN'S DISCOVERY OF X-RAYS

Working late in his laboratory one evening in 1895, a competent (but not very famous) scientist named Wilhelm Roentgen made a sensational discovery. His experiments revealed the existence of a new kind of ray that had exotic and interesting properties. Because these mysterious rays were then unknown, Roentgen called them x-rays (x standing for the unknown), a name that we still use to this day. After he reported his new discovery, Roentgen immediately became a highly celebrated figure and won the first Nobel Prize in physics just a few years later.

Of course, we now know what x-rays are. X-rays are similar to light, radio waves, infrared and ultraviolet rays, and a variety of other such radiations. All of these things are particular kinds of electromagnetic waves, so called because they are wavelike transmissions in electric and magnetic fields. The major difference between light and x-rays (and all the other types) is the wavelength of the radiation (this is the distance over which the wave repeats itself; different colors of light also differ in wavelength). The energy of the radiation also changes with the wavelength. X-rays have hundreds of times more energy than light, which accounts for both their usefulness and also their potential danger. This high energy also played an important role in Roentgen's discovery.

The experiments that Roentgen had in mind built on the work of many other nineteenth-century scientists (Thomson, Crookes, Lenard, and others). This work consisted of experiments with something called a cathode ray tube. These devices are not as unfamiliar as you may think; the picture tube in your television is a cathode ray tube. Basically, a cathode ray tube is just an airtight glass container with all the air pumped out to create a vacuum inside, and pieces of metal sealed into the glass wall so that electrical connections outside the tube can produce voltages on the metal inside the tube. If the voltage is high enough, a beam of electrons leaving the metal can be produced. A substance that glows when high-energy rays strike it, called a phosphor, can also be placed inside the tube. When the beam of electrons strikes the phosphor, we can see the presence of the beam by the telltale glow emitted. In essence, this is how your television creates the picture you see on the screen.

In 1895, the existence of electrons was not known (Thomson was soon to discover the electron in 1897). The cathode rays, which we now call electron beams, were at that time simply another mysterious radiation that scientists were still investigating. One important property known to be true of the cathode rays is that they are not very penetrating, that is, do not go through matter easily. For example, cathode rays couldn't escape through the glass walls of the tube. Lenard had discovered that a thin aluminum sheet covering a hole in the glass allows the cathode rays through, but the rays can then only make it through about an inch of air. All these observations were made using the glow of phosphors to detect the presence of the beam. Roentgen wondered whether some tiny portion of the cathode rays might after all be escaping through the glass walls undetected. The glass itself is weakly luminescent when struck by cathode rays, so the whole tube produces a kind of background glow. If an escaping beam were very weak, the slight glow it caused on a detecting phosphor might be washed out by this background glow of the tube. So Roentgen designed an experiment to test this hypothesis. He covered the tube with black cardboard to screen out the background glow, and his plan was to look for a weak glow on the phosphor he used as a detector when he brought it close to the covered tube wall.

As a first step, Roentgen needed to check his cardboard covering to make sure that no stray light escaped. As he turned on the high voltage, he noticed a slight glimmering, out of the corner of his eye, coming from the other side of his workbench (several feet away from the tube). At first, he thought that this must be a reflection from some stray light that he had not managed to block successfully. But when he examined the source of the glimmer more carefully, he was shocked to discover that it was coming from a faint glow of the phosphor he planned to use later as a detector. Something coming from the tube was causing a slight glow from a phosphor located over thirty times as far away as cathode rays can travel through air. Roentgen immediately realized that he had discovered some fundamentally new kind of ray, and he excitedly embarked upon the task of studying its properties. He found that these rays had extremely high penetrating powers. His phosphor continued to glow when a thousand page book or a thick wooden board was placed between the tube and the phosphor. Even thick plates of metals such as aluminum and copper failed to stop the rays completely (although heavy metals such as lead and platinum did block them). In addition to their penetrating power, Roentgen found that his new rays were not affected by magnetic and electric fields (in contrast to cathode rays, which are deflected by such fields).

In the course of his investigations, Roentgen made another accidental discovery that insured his fame in the history both of physics and of medicine. While holding a small lead disk between the phosphor screen and cathode ray tube, Roentgen observed on the screen not only the shadow of the disk but also the shadow of the bones within his hand! Perhaps to convince himself that the eerie image was truly there, Roentgen used photographic film to make a permanent record. After he completed his systematic and methodical investigations of the properties of x-rays, Roentgen published a report of his findings. The experiments were quickly replicated and justly celebrated. In physics, the discovery of x-rays opened up whole new avenues in the investigations of atoms and turned out to be the first of several revolutionary discoveries (followed quickly by radioactivity, the electron, the nucleus, etc.). In medicine, practitioners quickly realized the diagnostic value of x-rays as a way to look inside the body without cutting it open. The use of x-rays in medicine is one of the fastest practical applications of a new scientific discovery on record.

Roentgen's discovery of x-rays was a marvelous combination of luck and skill. Discovering something you aren't looking for, a process often referred to as serendipity, is not uncommon in the sciences. But as Pasteur's famous maxim says, "chance favors only the prepared mind." Roentgen's mind was extremely well prepared to make this discovery, both by his skill in experimental techniques and by his thorough knowledge of the previous work on cathode ray phenomena. Also, Roentgen's painstaking detailed investigation of the x-rays, following his initial lucky break, was crucial to the discovery process. He recognized the importance of the faint glimmer he did not expect to see.

§2. Detailed Background and Dreamlike Vision: Kekulé's Discovery of the Structure of Benzene

The carbon atom has chemical properties that set it apart from all other elements. Carbon is able to form a wide variety of chemical bonds with other elements, particularly with hydrogen, oxygen, nitrogen, and with other carbon atoms. The tendency to form various kinds of carbon-carbon bonds, in addition to the C-H, C-O, and C-N bonds, fosters the creation of complicated chainlike structures in such carbon-based molecules. For these reasons, many thousands of these carbon compounds exist, so many in fact that the study of them is a separate branch of chemistry. This branch is called organic chemistry, because it was once thought that only living organisms could produce these compounds. It's true that the molecules of living organisms (carbohydrates, fats, proteins) are all in this category, but "organic" is a misnomer in the sense that many organic chemistry compounds have nothing at all to do with life.

We might say that organic chemistry started with the synthesis of urea in 1828 by F. Wöhler. For many years thereafter, organic chemistry proceeded by trial and error, with chemists using their experience and various rules of thumb to synthesize new compounds. Organic chemists had no theory underlying their work and didn't know the structures of the compounds they created. Around the middle of the nineteenth century, the work of many chemists contributed to a growing understanding of the science underlying organic reactions and syntheses. Prominent among these chemists was August Kekulé. Kekulé's major contribution to organic chemistry was the idea that a molecule's three-dimensional structure was a key ingredient in determining that molecule's properties. The number of atoms of each element making up the molecule is obviously important, but how they are connected to each other in space is equally important. Kekulé's theories concerning molecular structure in general, along with his determinations of the structures of many specific compounds, advanced the field considerably.

By 1865, Kekulé had worked out the structures of many compounds, but the structure of benzene had proven to be intractable. Benzene is a volatile liquid that can be obtained from coal tar. Benzene is sometimes used as an industrial solvent, but the major importance of benzene is its role as the structural basis for many dyes, drugs, and other important chemicals. Michael Faraday had already determined the atomic composition of benzene in 1825. Benzene consists simply of six carbon atoms and six hydrogen atoms. But forming these six C and six H atoms into a structure that makes sense had defied the efforts of organic chemists, including Kekulé. One major problem with devising a reasonable benzene structure is the 1:1 ratio of C atoms to H atoms. Kekulé had already previously concluded that C atoms make four bonds to other atoms and that H atoms make one such bond, a system that works well for methane (see chapter 18) and similar compounds. But it's hard to reconcile this idea with the 1:1 ratio of C atoms to H atoms in benzene. Another big problem was the chemical behavior of benzene, especially compared to other compounds in which hydrogen atoms don't use up all of the available carbon bonds. These other compounds, such as acetylene (the gas used in welding torches), can be chemically reacted with hydrogen to produce new compounds that have more H atoms. Benzene, however, wouldn't accept any new H atoms in such a reaction.

Kekulé had pondered these problems for a long time. He combed his knowledge of organic chemistry in general, reviewed everything that was known about the reactions of benzene with other chemicals, and expended great effort in order to devise a suitable structure that made sense. Then, Kekulé hit upon the answer in a flash of inspiration. As Kekulé recounts the episode:

> I turned my chair to the fire and dozed. Again the atoms were gamboling before my eyes. . . . My mental eye, rendered more acute by repeated visions of this kind, could now distinguish larger structures of manifold conformation: long rows sometimes more closely fitted together all twining and twisting in snake-like motion. But look! What was that? One of the snakes had seized hold of its own tail, and the form whirled mockingly before my eyes. As if by a flash of lightning I awoke; and this time also I spent the rest of the night in working out the consequences of the hypothesis.

Kekulé's vision had suggested to him the ring structure of benzene shown in Figure 1. By having the chain of carbon atoms close on itself, he was able to satisfy the bonding numbers for C and H while leaving no room for additional H atoms. The question then became purely empirical. Does this benzene structure explain all of the known reactions and syntheses involving benzene? Does it predict new reactions and syntheses accurately? To make a long story short, the answer to these questions turned out to be, basically, yes.

Other structures were also proposed for benzene, and a vigorous debate went on for some years. In the end, Kekulé's ring structure had the most

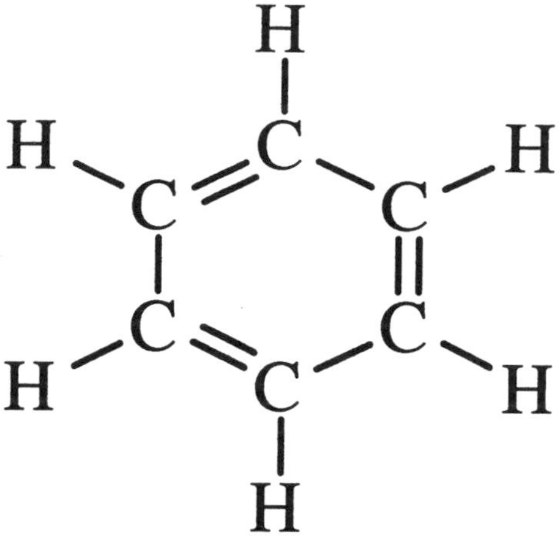

Figure 1. The structural model of the benzene molecule worked out by Kekulé, often referred to as a benzene ring. The ring structure was inspired by Kekulé's vision of a snakelike chain of atoms closing on itself.

success in explaining the data and became accepted as the correct structure. Some inconsistencies remained; calculated energies for the molecule were higher than the measured energies, and the placement of the three double bonds was distressingly arbitrary. These problems were finally cleared up many decades later when the modern quantum theory of chemical bonding was applied to the benzene ring, showing that all six bonds are really identical (circulating clouds of electrons bonding the carbons might be a more appropriate image than alternating double and single bonds). Meanwhile, Kekulé's proposed benzene ring was extremely successful in suggesting reaction pathways for commercially important organic compounds. The German chemical industry soon became the envy of the world, producing dyes, drugs, perfumes, fuels, and so on. The solution of the benzene structure problem was a key to much of this activity, which was an important segment of the German economy prior to World War I. Kekulé himself, however, had little interest in commercial ventures and confined his attention largely to scientific understanding.

A number of scientists have reported experiences similar to that of Kekulé. After a prolonged period of apparently fruitless concentration on a problem, the solution seems to arrive all at once during a brief period of relaxation. It's crucial to immerse oneself completely in the details of the problem before the flash of inspiration can come. An unusual aspect of Kekulé's experience is the highly visual character of his insight. His earlier development of the structural theory of organic chemistry had also been informed by such visions of dancing atoms, so this seems to have been a general part of his thinking process. Kekulé's early training had been in architecture, and it's possible that this training influenced his rather visual approach to chemistry and his tendency to think in terms of the spatial "architecture" of molecules.

§3. IDEALIZED MODELS AND MATHEMATICAL CALCULATIONS: THE DISCOVERY OF BAND STRUCTURE IN SOLIDS

Semiconductors are now an essential part of modern life, forming the heart of integrated circuits and diode lasers. Computers, compact discs, telecommunications, audio amplifiers, television, and many other devices would not exist if we didn't understand the behavior of semiconductors. The essential concept needed to understand semiconductor behavior is the concept of energy bands separated by band gaps, although few people have ever heard these terms. The existence of energy bands in solid materials was discovered by several people during the years from 1928 to 1931, at a time when semiconductors were merely a laboratory curiosity of little or no interest to anyone. The motivation for the work that led to this discovery was a desire to understand how electrons can even move through metals at all. If you imagine the negative electrons in a metal as moving through the array of fixed positive ions (which are much more massive than the electrons), the problem becomes apparent. The electrons and ions exert strong forces on each other. As the electrons try to move, they soon collide with an ion and are scattered into a different direction. This kind of scattering, in fact, is what causes electrical resistance in the first place. However, all the calculations done before 1928 indicated that the electrons shouldn't get much farther than one or two ions; experimental resistance measurements required electrons to get past hundreds of ions before colliding. This was a mystery.

In an effort to solve this mystery, Felix Bloch applied the newly invented theory called quantum mechanics to the problem. In the strange world of quantum mechanics, the electrons may be pictured as waves rather than as particles. Bloch also used another recently discovered fact: the ions in a metal are arranged in an orderly periodic fashion (a crystal lattice; see chapter 18). So Bloch's model (see chapter 6) of a metal consisted of quantum mechanical electron waves traveling through a periodic lattice of positive ions. Bloch succeeded; he was able to calculate the motion of the electrons in such a system, and the results were remarkable. It turned out that the electrons could sail effortlessly through the lattice without hitting ions. Resistance was due to vibrations of the ions and imperfections in the crystal. The results agreed well with experiments.

Another important step was taken by Rudolf Peierls, building on the foundation of Bloch's work. Peierls kept the same basic model that Bloch used, but now he varied the strength of the forces between the electrons and the ions. In his previous work, Peierls had already shown that a more detailed examination of Bloch's calculations reveals a "flattening" of the energy curve for the electrons. (This energy curve tells us how the electron's energy changes as its momentum increases.) His experience with this previous work enabled Peierls to recognize the significance of his new calculations. He discovered that where the flattening of the energy curve ends, there is an energy range above it in which no electron states at all can exist. Above this range of forbidden energies, another allowed band of electron energy states occurs. In other words, we have bands of allowed energy states for electrons in solids, separated by a zone of forbidden energies with no states. This zone of forbidden energies is what we now call a band gap. Above the second band of states, there is another gap, and so on. This discovery was an unexpected result of the calculations, and its importance can't be overemphasized. The idea of energy bands and gaps is at the heart of our understanding of the behavior of electrons in solids, but even Peierls did not see this clearly at first. One more ingredient was needed in order to fully appreciate the true significance of band structure in solids.

This final ingredient was supplied by Alan Wilson in 1931. The problem that Wilson was pondering concerned a peculiar implication of the work done by Bloch and Peierls. If electrons can move easily through a lattice of ions, no matter how strong or weak the electron-ion forces are, *then why isn't every solid a metal?* While grappling with this puzzle, Wilson realized that the proper interpretation of the band calculations not only answers the question, but does so in a fundamental and illuminating fashion. Wilson realized that if a band was full (all possible states occupied by electrons), then no electron could gain energy, because to gain energy puts the electron into the band gap where no states exist for it to occupy. Electrons must gain some energy to become a current (i.e., to move), as in a conductor. A solid with a full energy band must then be an electrical insulator (like quartz or sapphire). Solids with partly empty bands have higher energy states available for the conduction electrons to go into, and so these are metals (like copper or aluminum). Wilson's idea explained the essential difference between metals and insulators, which had been an unsolved problem since the first attempts to understand the properties of matter.

Going further, Wilson extended his theory to explain electrical conduction in semiconductors (like silicon). The major riddle presented by semiconductors was that they, in contrast to metals, became *better* conductors at higher temperatures instead of worse. Wilson pictured semiconductors as solids with full bands but having rather small band gaps. Electrons can be thermally excited into the empty band above the gap, and these electrons conduct the current. Naturally, more electrons can acquire enough energy to cross the gap at higher temperatures, and so the sample becomes a better conductor. Using the idea of energy bands and gaps, we could now understand the electrical behavior of metals, insulators, and semiconductors in a unified manner. At the time, experimentalists were still debating whether semiconductor behavior was real or just an artifact caused by low-quality samples. Several decades later, our understanding of semiconductors became the basis for the microelectronics revolution. The central unifying idea of energy band structure in solids arose unexpectedly from the results of calculations. No one anticipated the existence of band gaps in solids (in fact, as we've seen, it took a while to recognize their importance even after the discovery). The concept just turned out to be a result of assuming electron waves in a periodic lattice of ions, and calculating the consequences of this assumption.

§4. Exploration and Observation: Alexander von Humboldt and the Biogeography of Ecosystems

Although he is not a famous figure today, Alexander von Humboldt was one of the leading natural scientists of his own time. He was a friend or correspondent to virtually every noted scientist in Europe, he socialized with the elite in the court of Napoleon, he was admired by Goethe, he stayed at Jefferson's home Monticello as an honored guest, and the King of Prussia put some effort towards attracting Humboldt into his service. Humboldt was probably more well known to his contemporaries in the educated public than any scientist alive now is known. His fame is reflected in the twenty-four places (towns, counties, mountains, rivers, even a glacier) named after him. Humboldt's scientific work is voluminous, and he worked in virtually every field in the natural sciences. He made contributions to astronomy (studying meteor showers), botany (discovering over three thousand new species), geology (studying volcanoes and geologic strata), geophysics (studying the earth's magnetic field), meteorology (studying tropical storms), and oceanography (studying the major ocean currents). And this list isn't even complete. Many of these studies were observational in nature; Humboldt's sharp mind, natural curiosity, and keen powers of observation gave him the intellectual tools needed for such work. But what also set Humboldt apart from the average naturalist was that he embarked on a voyage of exploration that can only be called epic. Humboldt was the scientist who opened up the New World for study.

In 1799, Humboldt embarked on his journey of exploration to South America and Mexico. When he left Latin America in 1804, he had collected thirty cases of geological and botanical specimens, as well as innumerable notes, records, measurements, maps, and codices. Among his adventures on this trip, he traveled by boat down the Orinoco River through the tropical jungles of Brazil (coming down with typhoid in the process). In Ecuador, he scaled the highest peak in the Andes, setting a new record for the highest altitude ever achieved. Everywhere he went, he made precise measurements of the latitude (using astronomical instruments), barometric pressure, and the earth's magnetic field (strength and direction). He collected gases from the fumes of active volcanoes and analyzed their chemical composition. He described the geological structures and climates of the different regions he visited. Everywhere he went, Humboldt collected samples of minerals, plants, and animals. In the mountains of the Andes, Humboldt made one of his most important and fruitful discoveries. As he climbed the mountains on exploratory expeditions, Humboldt was struck by the dramatic changes in the vegetation and animal life at different elevations. At the base of the mountains grew palms, bananas, and sugar, typical of the tropical climate. At higher elevations, coffee and cotton were found, along with maize, wheat, and barley on the flatter areas. Above this, the vegetation became more sparse, mostly evergreen shrubs, while at the highest elevations only alpine grasses and lichens could be seen. He realized, based on his extensive travels, that this sequence was similar to the changes in vegetation with latitude as one moved from the tropics towards the poles. As Humboldt pondered the meaning of these changes, he realized that the climate was a major, but not the only, part of the physical environment that determined the plant life found in a geographical area.

Based on his studies and observations, Humboldt developed a theory of biogeography, of how the physical conditions of an area influence the features of the ecosystem (to use the term we now employ) found there. The temperature, soil conditions, amount of sunlight, rainfall, and topography all work together to determine what kind of plant and animal life might inhabit a place. This may seem obvious today, but the idea was both novel and important when Humboldt proposed and explicated it. Much of his work consisted of describing and classifying parts of nature, but this theory gave meaning and context to the classifications. Humboldt believed in the underlying unity of nature, and in the biogeography idea he could see a reflection of this unity. Like any important discovery, his idea also opened up new areas of investigation and suggested new ideas to other scientists who followed Humboldt.

Before leaving Humboldt, I can't resist the temptation to mention his work in cultural anthropology. The native cultures of the Inca, Aztecs, and Maya had been partially decimated by conquest, but a keen and intelligent observer like Humboldt was still able to learn and record a great deal. He studied their languages, visited archeological sites (such as the pyramids at Teotihuacan), collected ancient writings and sculptures, recorded their myths and legends, and examined petroglyphs. The knowledge of astronomy possessed by the vanished cultures was an especially interesting area studied by Humboldt, and he looked in detail at the calendar systems that they had created. These cultures had mostly been ignored before Humboldt's work, and his efforts stimulated further interest by later scholars. Humboldt's insatiable curiosity and sharp analytical mind ranged over every part of the natural world. His travels and explorations gave him the opportunity to deliver a treasure trove of new knowledge to the intellectual community of Europe, and this knowledge contributed to the great integrative theoretical work in geology and biology done by scholars who followed him in succeeding generations. Humboldt's own attempt at a grand integration of all knowledge was his masterpiece, *Cosmos*. This work is informed by Humboldt's conviction of the harmony and unity underlying the diversity of nature. Subsequent discoveries and theories have rendered the details of *Cosmos* obsolete, but it remains a remarkable testament to the depth of Humboldt's thinking.

§5. The Hypothetico-Deductive Method: Edward Jenner and the Discovery of Smallpox Vaccine

Edward Jenner started his career as an apprentice country doctor in Gloucestershire, the rural area of England where he grew up. After his apprenticeship, Jenner went to London in 1770 for more advanced training under the highly regarded surgeon, John Hunter. Medicine was still in a somewhat primitive state at this time, using many traditional methods of doubtful efficacy. Hunter was a pioneer in the application of scientific thinking to medical practice, and he taught Jenner to do the same. Jenner proved to be an excellent student, and he developed into a first rate doctor under the guidance of the brilliant Hunter. Equally able as both a medical practitioner and as a scientist, Jenner embarked on his own career after his time with Hunter ended.

Against his teacher's wishes, Jenner decided to move back to Gloucestershire (Hunter wanted Jenner to stay in London, where he could make a reputation). His move back to the countryside, however, gave Jenner the opportunity to follow up on an idea he had gotten when he was still a young apprentice. While treating a milkmaid for a minor ailment known as cowpox, Jenner had become acquainted with one of the local legends of the Gloucestershire region. The milkmaid told him that she was lucky to have the cowpox, because now she would never contract smallpox. Smallpox was one of the most dreaded diseases of the time, and the milkmaid assured Jenner that having had cowpox protects against getting smallpox. Jenner put this conversation in the back of his mind at the time, but now he was hoping to look into the matter more thoroughly. Gloucestershire was a major dairy farming area, and this made it an ideal place to conduct his study. Cowpox was a disease that the cows contracted, and the cows often then transmitted the illness to humans (through cuts on their hands, for example) as they milked the cows. Cowpox wasn't very serious; it caused fever, aching, and some temporary blisters around the hands. The illness lasted a few days, and a full recovery could be expected. The cowpox sometimes came to the dairies in epidemics, but sometimes it vanished for years on end.

Smallpox at that time was a worldwide scourge, highly contagious and often fatal. In the century before Jenner began his work, smallpox had claimed over twenty million people in Europe. Almost a third of the children under age three in Britain succumbed to the Red Death. The smallpox sometimes raged unchecked in terrible epidemics, and there was no treatment available. Among the victims who did not die, many were left horribly disfigured, blinded, or insane. The only preventative measure known was inoculation, the practice of purposely infecting people with material from active smallpox pustules. This action might produce a less severe case of the illness, which then protects the person against contracting it again. But, the procedure often could go awry and produce a severe case, even death. Worse yet, even when the procedure worked well, the inoculated person might give the disease to others. In Russia, an entire epidemic had started this way. So Jenner's idea that there might be a safe way to prevent smallpox was exciting. It had taken root in his mind and become his dream: a world free from the Red Death. But the matter was not simple. There were a number of cases in which people who had once had cowpox did come down with the smallpox. For that reason, many of the local Gloucestershire doctors dismissed the old legend completely. And yet, there was enough anecdotal evidence in favor of the legend to still convince many people of its truth. Jenner, excellent scientist that he was, realized that he needed to start making careful observations, including keeping good notes and records, if he wished to untangle the situation.

After his medical practice was established, Jenner began his work in earnest. He made a scientific study of the cowpox, which no one had ever done before. A more precise description of the symptoms and course of the disease was needed, both in cows and in humans. For several years, Jenner carefully observed all the cases which occurred in the dairies, and he interviewed people who had gotten cowpox in the past. Making careful notes on these case histories, he began to achieve a more thorough understanding of the cowpox. At the same time, Jenner began to make a systematic study of the cases in which cowpox had apparently conferred immunity to smallpox. Just as importantly, he also studied those cases where it had not done so. If he wished to use cowpox as a tool in the fight against smallpox, Jenner would have to solve the puzzle of why some people still contract smallpox even after having cowpox. These cases were often taken to be proof that the old legend was merely superstition, and Jenner's work was widely regarded by his colleagues as a waste of time.

But Jenner did not give up easily, and he continued to look for some clue that would solve this puzzle. As he pored over his records, he came to realize that different sets of symptoms were observed (the appearance of the pustules, swelling in the armpits, headaches, body pains, vomiting, etc.) in different victims; in other words, the cowpox had no single fixed description. The same thing was true in cows (sometimes the pustules were circular, sometimes irregular; sometimes they lasted weeks, sometimes days, and so on). Jenner concluded that what dairy farmers had been calling cowpox was actually several distinctly different diseases. This fact solved Jenner's puzzle, because only *one* of these diseases conferred immunity to the smallpox. Once he had this idea with which to organize his observations, Jenner was soon able to distinguish these different versions from each other. His next step was to determine which disease (he referred to it as the true cowpox) was able to protect against smallpox. Based on his records and observations, Jenner was able to give a very complete and accurate description of the true cowpox. One of the major clues that helped him was the lack of symptoms in response to inoculation with smallpox matter on the part of people who had contracted the true cowpox in the past. In this manner, after five years of patient work, Jenner was able to distill a hypothesis from his observations.

He then put his hypothesis to the test, and he discovered that some mysteries still remained. At one of the local dairies, there was a major outbreak of the true cowpox. Jenner continued to keep his meticulous records, and so there was no doubt in his mind when these same milkers came down with smallpox the following year. His hypothesis, and his dream of defeating smallpox, seemed shattered. He pored over his records and continued to study the cowpox, looking for a solution to this new puzzle. For several more years, Jenner tried in vain to figure out why even the true cowpox sometimes failed to protect against smallpox. There seemed to be no answer. Then, while looking at two cows in different stages of the disease, Jenner realized the factor he had been overlooking for so long. The disease, and in particular the appearance of the pustules, gains in strength for a few days, then the disease is at its worst for a while, and finally it declines and goes away over a few days. This much Jenner had known well for years. But now he hypothesized that the virulence of the matter in the pustules, which transmits the disease, should also likewise gain and decline in strength; *and cowpox only protects against smallpox when the matter in the pustules is at its strongest.* This new hypothesis solved the puzzle, and was consistent with the facts he knew. For example, the milkers whose smallpox epidemic so mystified him had gotten cowpox in its earliest stages.

Jenner now designed an experiment to test this latest hypothesis. In May of 1796, Jenner extracted some material from a pustule on the hands of a milkmaid named Sarah Nelmes. She had contracted the disease from a cow while it was at its worst, and her own case was also now at its strongest. These conditions were ideal for Jenner's experiment, and he used the material to purposely infect a young boy named James Phipps. After the cowpox had run its course in young Phipps, Jenner inoculated him with live smallpox matter the following July. Tensely, day after day, Jenner and the Phipps family looked for any sign of a smallpox infection beginning. But even several days after the expected time of onset, the boy had absolutely no smallpox symptoms! The experiment had succeeded. The cowpox material, deliberately introduced into a human body, had been shown to confer immunity to the Red Death.

Our story ends here, but Edward Jenner's story went on for several more years. He had an uphill battle convincing the medical community and the general public that his method, which came to be known as vaccination (from the Latin word for cow), was an effective method to prevent smallpox. His problems were compounded by incompetent people who tried to steal his idea but couldn't perform the procedures properly (in one terrible case, a quack mixed up cowpox and smallpox material, actually starting an epidemic). Such mishaps gave the vaccine an undeserved reputation for being unsafe and ineffectual. But Jenner managed to sort these problems out, and in the end his smallpox vaccine derived from the cowpox material came into widespread use, saving untold numbers of people from the ravages of smallpox. For this service to humanity, Jenner became a hero in his own time and had numerous honors bestowed on him.

There are certainly elements of luck and inspiration in this story, but it mainly illustrates the pathway to discovery that we now often call the hypothetico-deductive method. We start by making observations; organize these observations into a hypothesis; test the hypothesis against further observations and modify it as needed; make predictions based on the modified hypothesis and design experiments to test our predictions. This highly successful methodology is also sometimes enshrined in elementary textbooks as the "scientific method." The discovery of the smallpox vaccine is a good example of just how powerful this method can be.

Although Jenner is most famous for discovering vaccination, he had a productive career as both a medical researcher and as a naturalist. In medicine, he deserves some credit for discovering the role of hardening of the arteries in causing heart attacks. As a naturalist, he made important studies of hibernation in animals, and he also discovered the cuckoo hatchling's habit of pushing fellow hatchlings out of the nest. In all of his work, Jenner's careful, accurate, and honest observations were always the foundation for his conclusions.

FOR FURTHER READING

Humboldt, by Helmut de Terra, Alfred A. Knopf, 1955.
Jenner and the Miracle of Vaccine, by Edward F. Dolan, Jr., Dodd, Mead, 1960.
Kekulé Centennial, edited by O. T. Benfey and R. F. Gould, The American Chemical Society, 1966.
Serendipity, by Royston M. Roberts, John Wiley & Sons, 1989.
Out of the Crystal Maze, edited by L. Hoddeson, E. Braun, J. Teichmann, and S. Weart, Oxford University Press, 1992.
Ideas in Chemistry, by David Knight, Rutgers University Press, 1992.
Remarkable Discoveries!, by Frank Ashall, Cambridge University Press, 1994.
"Wilhelm Conrad Roentgen and the Glimmer of Light," by H. H. Seliger, *Physics Today*, vol. 48, no. 11, p. 25, November 1995.

EPIGRAPH REFERENCE: Michael Polanyi, *Science, Faith, and Society*, University of Chicago Press, 1964, p. 22.

Chapter 2

NATURE'S JIGSAW: LOOKING FOR PATTERNS
AS A KEY TO DISCOVERY

> ... merely to observe is not enough. We must use our observations and to do that we must generalize. .... The scientist must set in order. Science is built up with facts, as a house is with stones. But a collection of facts is no more a science than a heap of stones is a house.
>
> *(Henri Poincaré)*

TO DISCOVER an underlying coherence and regularity in nature, buried within reams of observational and experimental data that has so far defied understanding, is at the heart of science. Finding such previously unseen patterns is one of the key processes of scientific discovery. In this chapter, we look at the stories of two highly important discoveries. Each story has its own interesting features, but they both have in common the finding of a pattern, like the pieces of a jigsaw puzzle falling into place once you see the picture they form.

§1. THE PERIODIC TABLE OF THE ELEMENTS

Our concept of an element, a chemical substance that cannot be broken down or changed into anything else, was unknown to antiquity. Nevertheless, a few of the substances we now recognize as elements (mostly metals like gold, iron, and copper) were known in ancient times, and the alchemists later isolated a few more (such as antimony and arsenic). During the seventeenth and eighteenth centuries, chemistry became established as an empirical science, but it still lacked a theory in the modern sense. Valuable discoveries were made, especially in the work with gases; hydrogen, oxygen, and nitrogen were all observed during this time, but their nature was not understood. Lavoisier was finally able to reconceptualize chemistry in a way that made these observations sensible. He introduced the idea of an element in its modern form and he enumerated the substances that he considered elements at that time. Lavoisier published his work in 1789, and chemistry progressed rapidly after this. Many new elements were isolated and identified, using ingenious new techniques.

The novel science of electrochemistry, made possible by the invention of the battery, played a major role in this work. By about 1850, the number of known elements had grown to roughly sixty, compared to only around twenty in Lavoisier's time. Thanks to the skillful and methodical work of many chemists, a great deal of information about the chemical and physical properties of these elements was available.

What kind of chemical and physical properties are interesting? Whether an element is a metal or a nonmetal; a solid, a liquid, or a gas; brittle or ductile; highly reactive or fairly inert; all these are basic questions. The density, melting point, boiling point, and crystal structure are all basic physical measurements. Chemically, the first interesting question is: with what other elements does this form a compound? What are the ratios of these elements in the compounds they form? Based on the answers to these questions, chemists were able to assign a valency to each element, a measure of how much is needed to form a compound. Potassium, for example, has a valence of one and forms a compound with chlorine in a 1:1 ratio, a compound with oxygen in a 2:1 ratio, a compound with phosphorus in a 3:1 ratio, and so on. Calcium, with a valence of two, forms a compound with chlorine in a 1:2 ratio, a compound with oxygen in a 1:1 ratio, and so on. If you multiply these two simple examples many times over, you begin to have a sense of the vast quantity of information that chemists had acquired by the middle of the nineteenth century.

As the properties of the elements were explored, patterns began to emerge. For example, several of the metals all had very similar chemical properties (highly reactive with a valence of one) and physical properties (low densities and low melting points); they were called the alkali metals. Other sets of elements with strikingly similar properties were also known (the halogen gasses, the alkaline earth metals). Other more complex patterns were seen, such as the similarity in crystal structures between analogous compounds of elements (e.g., the cubic structure of salts like NaCl, KCl, LiF, NaI, etc.). The formation and reactions of acids and bases, carbonates, sulfates, and others could all be systematized based on the patterns seen in elemental properties. But there was no underlying concept to tie all of these patterns together coherently. The underlying concept turned out to be related to the atomic weights of the elements. (The atomic weight is a measure of the relative mass of an element's atom, based on some agreed-upon standard.) The nineteenth century chemists realized that the atomic weight of an element is a fundamentally important quantity, and they spent considerable effort to measure it well. Such measurements were very difficult, however, and the tabulated atomic weights remained very uncertain in many cases even toward the end of the century.

Relationships between chemical properties and atomic weights were noted as early as 1817, when a German chemist named Dobereiner devised a system of "triads." He observed that for various sets of chemically similar elements, the atomic weight of one is the average of the atomic weights of the other two (e.g., Ca is the average of Mg and Sr). In the following decades, a variety of relationships between chemical properties and atomic weights were proposed. Between 1862 and 1870, six different scientists devised periodic systems that were similar in conception to that used today. The most famous of these scientists, who is usually credited with the discovery of the periodic table of the elements, was Dmitri Mendeleev. The periodicity of the relationship between the atomic weights and the properties of elements was strongly hinted at by the known properties of the lighter elements, which seemed to recur in a cycle of eight (John Newlands in England referred to this in 1865 as the law of octaves). But not all of the elements fit into such a system; given the presence of discrepancies, many skeptics attributed the seeming pattern to coincidence. What set Mendeleev apart from the other investigators of this problem was his dogged determination to have a system that really worked and his vast detailed knowledge of the chemical and physical properties of virtually every known element.

Many of those properties he measured himself. Other properties he found by combing through the chemistry literature, always keeping up-to-date with the very latest work. In addition to reading the published work, Mendeleev kept up a lively correspondence with other chemists in order to make sure that he had the very best information about the elements. He was especially scrupulous about having the most accurate atomic weights possible, since these weights were the key variable in the periodic system. For Mendeleev, every single element was like a close friend whom he knew well. All of the information he gathered was written down on a little white card, one card for each element. Each new chemical fact he learned was added to the appropriate card. As Mendeleev slowly worked out the correct pattern for the variation of the elements' properties, he hung the cards on his wall in the proper order, moving them around as he got new ideas and updated information. Slowly, the patterns came to make more and more sense, and Mendeleev was able to apprehend a unified order amidst the multiplicity of elemental properties; he announced his periodic system of the elements in 1869. As the pieces of the puzzle fell into place and Mendeleev was able to fit most of the elements into his system, he became more confident. Because his work was grounded in an intimate knowledge of the elements, which included many thousands of pieces of factual information, this work was built on a sturdy foundation. After many years of work, the pattern seemed to be authentic and the system seemed complete barring a few exceptions. Mendeleev then became extremely confident. He became so confident, in fact, that he took an unprecedented action: Mendeleev claimed that the accepted atomic weights of several elements were wrong because they did not fit into his periodic system. He proposed changing the atomic weights of indium, uranium, cerium, and titanium so that these elements would fall into their proper place in the pattern. These changes weren't made in order to cheat by brushing exceptions under the rug; on the contrary, Mendeleev proposed these new atomic weights as bold predictions of his system, which could be experimentally tested to verify or refute the periodic law. After many careful measurements had been made, Mendeleev's predictions were in fact largely verified.

Mendeleev made another set of predictions that were even more dramatic. His periodic table contained a number of blank spaces. Mendeleev declared that these blank spaces must correspond to elements as yet undiscovered. By using the periodic table, the thorough Russian chemist was able to predict the atomic weights of the unknown elements and to provide a long detailed list of the chemical and physical properties that these predicted elements would possess. The new (still undiscovered) elements were named after the elements with analogous properties above them in the table, for example ekasilicon and ekaaluminum. Just a few years later (in 1875), a new element was discovered in France with an atomic weight and set of properties matching the predictions for ekaaluminum; we now know this element as gallium, named after its country of origin. Another new element, named germanium, was later discovered and matched with ekasilicon in atomic weight, physical properties, and chemistry. Few could doubt the correctness of the periodic law after this. Mendeleev lived to see a number of his predictions verified. Another stunning verification of the system, however, could not have been predicted. When the first inert gas was discovered (argon; see chapter 4), it didn't fit anywhere in the periodic system. But a whole set of such inert gases were soon found, and they filled up a whole new column placed between the halogens and the alkali metals, with atomic weights that were just right to keep the period system intact. The most recent additions to the periodic table are the short-lived artificial elements of very high atomic weight, one of which is named mendelevium in honor of the discoverer of the periodic law.

Actually, a number of people discovered some version of the periodic law at roughly the same time. New versions have also been devised since then. An explanation for the amazing regularity in the elements was found after Mendeleev died by Bohr, Pauli, and others (see chapter 18). Mendeleev was not the first to note these regularities and did not explain them. What Mendeleev is justly honored for is the discovery of a pattern incorporating such a wealth of factual detail that it had to be real.

§2. DRIFTING CONTINENTS

If you look at a world map or globe, you may be struck by how the coastlines of Africa and Europe (on the eastern side of the Atlantic Ocean) seem to match up with the coastline of the Americas (on the western side). It's not a perfect fit, of course, but the correspondence of these coastlines is striking enough to have fired the imagination of several writers. A number of people before the twentieth century had speculated that these continents might have once been joined together. In 1910, this thought crossed the mind of Alfred Wegener. The major difference between Wegener and the other people who had noticed this fit between the continents is that Wegener looked into the matter more closely. He studied the geology of the African and South American coastal regions, and learned about the plants and animals living there. Paleontologists had collected a lot of information about the fossils of plants and animals that had once lived on these coasts, and Wegener also studied this fossil record. Much of the information fit together like the pieces of a puzzle, and Wegener became convinced that these continents had once been joined together, 250 million years ago. Since then, they have been drifting apart to their present positions. It turns out that similar correspondences exist between the coasts of east Africa and India.

The idea of continental drift was not accepted by very many people when Wegener proposed it. After all, the thought of continents drifting around like icebergs is rather absurd by common sense standards, and probably aroused opposition for that reason alone. But even on strictly scientific grounds (logic and evidence), there were good reasons not to believe. An alternate explanation for the similarities in geology, fossil record, and plant/animal life had already been advanced and was widely held. This alternate theory assumed that land bridges connected the continents in the past, and these bridges have now sunk under the oceans. Meanwhile, there was a major problem with the drift theory: no mechanism was known that could account for the huge forces that might cause continents to move. Wegener had in fact proposed such a mechanism, but it was easy to show that his proposal must be wrong. He speculated that small differences in gravity between poles and equator might be combined with tidal forces to move the continents. These forces, however, are millions of times too small for this job. Critics of continental drift, ignoring the vast amount of empirical support Wegener had presented, dismissed the theory because his mechanism seemed so clearly wrong.

Wegener admitted that his mechanism was speculative, but he insisted that the evidence in favor of moving continents was strong (whatever the mechanism may turn out to be). Sinking land masses (in other words, the proposed former bridges) made no sense to him. Land masses are higher (continents) because they are less dense or lower (ocean floors) because they are more dense. Large land mass areas wouldn't sink or rise at random. Also, certain life forms were found only in a *narrow* range *near* both coasts. Surely detached and drifting continents explained this fact better than a vast former land bridge. Finally, Wegener's theory explained why the climates of the continents had been so very different in the distant past (namely, these continents had been located at very different places on the earth's surface then). The opponents of drift also had some good arguments. For example, the lower crust of the earth is not fluid enough to allow the upper crust to move. In their judgement, the geological and paleontological evidence was too fragmentary and incomplete to prove the case for drift. These opponents also charged that many of Wegener's coastal fits were not as good as he claimed (the opponents were mistaken in this case; Wegener had quite rightly used the boundaries of the continental shelf rather than the coastline itself).

As you can see, both sides of the controversy had some good arguments. The pieces of the puzzle could be fit together in more than one way, and there was no definite choice as to which picture was correct. Wegener had a few influential allies in the scientific world, but his opponents were both influential and also far more numerous. Wegener's theory was a radical innovation, and the majority of people in the scientific community were not prepared to accept it without decisive proof. More pieces of the puzzle were needed.

One major missing piece, obviously, was a credible mechanism to drive the drift process. Such a mechanism was actually proposed as early as 1928 by Arthur Holmes: convection currents (see chapter 17) in the earth's mantle. The idea is based on a well-known fact, namely, that the earth's interior is continually heated by radioactivity. This heat must somehow escape from deep inside the earth. Holmes proposed that the heat moves upward in a convection current (i.e., fluid mantle material carrying the heat as it moves) that then flows sideways along the boundary with the crust and eventually flows back downward (where it heats up again to renew the cycle). As the mantle convection current moves along under the crust, the crust is carried along with it like a ship carried by a water current. Interestingly, this work did not attract much support for the idea of continental drift. Perhaps the concept had been too thoroughly dismissed by that time for anyone to think seriously about it. And of course, the convection currents themselves were still unproven ideas. Few scientists are willing to give up lightly the ideas on which their entire careers are based. In any event, Wegener had little support for his theory when he died in 1930 during an expedition to Greenland. The new pieces of the puzzle needed to support drift theory would be empirical, not conceptual. These pieces would not be found for several decades, in unexplored territory at the bottom of the oceans.

Almost nothing was known about the ocean floor before 1940, with one exception. A major mountain had been discovered in the middle of the Atlantic Ocean (the Mid-Atlantic Ridge) when the first telegraph cables were strung from America to Europe. After World War II, a great deal of effort was expended using newly available technologies (like sonar, deep-sea core sampling, and seismography) to learn about the bottom of the sea. The result of this research revolutionized our thinking. The Mid-Atlantic Ridge turned out to be part of a worldwide system of undersea mountain chains. Remarkably, these mountains have at their heart a huge chasm, about one mile deep and twenty miles wide. This rift is a major source of seismic and volcanic activity. Another amazing discovery was that the ocean floors are quite young (by geological standards, anyway). No fossils or rocks older than a few hundred million years could be found. This may sound old, but the long-held idea of geologists was that the ocean floor must be among the oldest places on earth, undisturbed for perhaps billions of years. The composition of the ocean floor was also unexpected, being mostly basaltic rocks in contrast to the mostly granitic rocks making up the continents. These surprising results were very difficult to explain.

An explanation that integrates all of these new facts was proposed around 1960 by Harry Hess. Termed "sea-floor spreading," the idea is that new material is continuously welling up (from the earth's interior) in the great oceanic rifts. This new rock is added to the ocean floor as it comes up, pushing the previously added rock outward from the rift. The sea floor is being newly created and spreading outward (from the rift) all the time. The engine driving this process is the mantle convection mechanism that we've already considered. Meanwhile, the continents ride along on the spreading sea floor like boxes on a conveyor belt. In this way, new ideas that were proposed in order to explain perplexing new oceanographic facts also implied the reality of continental drift. At the same time these new explorations of the ocean floor were going on, other geologists were engaged in studying a very different field: paleomagnetism. The basic idea is simple. If lava contains magnetic minerals, those minerals will line up with the earth's magnetic field like a compass needle to point toward the magnetic North Pole. As the lava cools and solidifies, these pointers will be frozen in place and keep their original direction forever. If you go to various locations on the earth's surface, you might then expect to find the magnetic orientation of the rocks pointing toward the North Pole. Geologists did indeed expect this result. Instead, they discovered a variety of different directions. This work is complicated, because you have to be sure the rock hasn't been moved (by earthquakes, water, etc.) since the lava cooled. Eventually, a great deal of work led to a consensus that the differences in direction were real. The perplexing conclusion seemed to be that the North Pole wanders around over geological time periods.

But there is another way to interpret this data. Perhaps it is not the Pole that is wandering; perhaps the continents are wandering. Based on this hypothesis that the continents are moving with respect to the North Pole, geologists compared the magnetic directions of the rocks at many geographic locations and time periods to figure out the motions that the continents must have had. The results were astounding. Based on this study of fossil magnetism, the continents must have moved in just the way that Wegener had claimed that they moved. Another piece of the pattern fell into place. Oceanography and paleomagnetism then joined forces to produce a striking confirmation of the picture that was emerging. Magnetic studies of rocks, on both the continental land masses and the ocean floor, revealed another initially confusing fact: a complete reversal of the magnetic direction for some rocks. This magnetic reversal is found to occur simultaneously at many places on the earth, and it has taken place more than once. In other words, the magnetic poles periodically flip places (this field reversal is a well-confirmed empirical fact, but we still don't fully understand why it happens). What does this magnetic reversal have to do with our story about continental drift?

The magnetic studies of the ocean floor, near the great oceanic rifts, revealed another strange pattern. Stripes of sea floor, hundreds of miles wide and running parallel to the rifts, contained rock with the same magnetic orientation. As you move out away from the rift, the stripes alternate in their orientation (imagine a map with North-pointing rock colored black and South-pointing rock colored white; this map will look a bit like a zebra). Once you realize that the earth's magnetic field periodically reverses itself, however, this seemingly bizarre result begins to make perfect sense. In fact, this result is a startling vindication of the theory of sea-floor spreading. The fresh lava emerges from the rift to create new ocean bottom. As the lava cools, it freezes in the direction of the earth's field. As time progresses, the new rock moves outward, creating one of the observed stripes. When the field reverses, creation of a new stripe begins. As the field reverses periodically, a pattern of alternating stripes is created. The theory predicts exactly what we observe. All of the puzzle pieces had now been found. The time was ripe for geologists to assemble all of these pieces into a coherent picture. This was accomplished in the middle of the 1960s by a number of people, the key player perhaps being Tuzo Wilson. The resulting picture, a synthesis of all available information, is what we now call plate tectonics. The continents ride on giant pieces of the earth's crust called plates. The plates themselves fit together on the earth's surface like the pieces of a puzzle. The bottoms of these plates reach down into the mantle to a region (called the asthenosphere), which is more plastic and fluidlike due to the high temperature. This region is where the great convection currents exist, slowly driving the movements of the plates. The oceanic rifts, the Pacific "ring of fire," and other unstable regions of high volcanic and seismic activity are plate boundaries where the mantle material is welling up or sinking down. The continental mountain ranges are the result of momentous collisions between the plates. Plate tectonics is now one of the fundamental organizing principles of modern geology.

FOR FURTHER READING

Famous Chemists, by W. A. Tilden, George Routledge & Sons, 1921.
Masterworks of Science, edited by J. W. Knedler, Jr., Doubleday, 1947.
Mendeleyev, by Daniel Q. Posin, McGraw-Hill (Whittlesey House), 1948.
The Periodic System of Chemical Elements, by J. W. van Spronsen, Elsevier, 1969.
Continents in Motion, by Walter Sullivan, McGraw-Hill, 1974.
Great Geological Controversies, by A. Hallam, Oxford University Press, 1989.
A History of Geology, by G. Gohau, Rutgers University Press, 1990.

EPIGRAPH REFERENCE: Henri Poincaré, *The Foundations of Science*, Science Press, 1913, p. 127.

Chapter 3

NEW VISTAS: EXPANDING OUR WORLD

WITH INSTRUMENTATION

> Not only do we use instruments to give us fineness of detail
> inaccessible to direct sense perception, but we also use them
> to extend qualitatively the range of our senses into regions
> where our senses no longer operate. . . .
> *(P. W. Bridgman)*

W E EXTEND our powers of observation by the use of instruments. We sometimes increase the range of our senses into new regimes of size or intensity, but we can also do more than just magnify our usual means of perceiving the world. By using appropriate instruments, we can even "observe" new phenomena that our senses can't detect at all. Chemists, for example, learn about the motions of atoms within a molecule by measuring the infrared rays that a molecule emits but our eyes can't see. Beyond extending the range of our senses, there is another way in which instruments help us discover new things. Instruments can also create new and exotic conditions under which to do experiments and make observations. (In technical jargon, instruments can extend the range of an independent variable). For example, the behavior of matter undergoes fascinating transformations at extremes (very high and very low) of temperature, pressure, energy, and so on. A famous particular case is graphite (pencil lead) turning into diamond at very high pressures and temperatures. In this chapter, we'll look in more detail at some examples of discoveries made by using new instrumentation to extend the range of our observations. It may happen that only marginal gains are made by such extensions; we learn a little more, increase our precision, tidy up a few details. But sometimes, dramatic new discoveries are made in these new territories and totally unsuspected phenomena emerge. The examples I've chosen mostly illustrate the latter, more dramatic, cases. Not only are such cases more interesting, but I think they are also more typical.

§1. SUPERCONDUCTIVITY

An important technological development that grew out of our understanding of thermodynamics (the science of heat; see chapter 17) during the nineteenth century was the invention of the refrigerator, the ability to make something colder. Along with its important practical uses, such as preserving food, refrigeration gave scientists a new tool in the study of nature. They could now explore the properties of matter at very cold temperatures. Nowhere was this tool more highly developed than in the laboratory of H. Kamerlingh Onnes in Leiden. Onnes was the leader of a large research group including master technicians and instrument makers as well as scientists, all devoted to the study of science at low temperatures (cryogenics). In 1908, Onnes and his colleagues made a major breakthrough: they achieved temperatures low enough to change helium from a gas into a liquid, just 4.2 centigrade degrees above absolute zero (absolute zero is the lowest temperature that can exist). Once they were able to make liquid helium, Onnes and his group could cool other samples to the same low temperature and study the behavior of these samples.

An important property that they wanted to study was the electrical resistance of metals. What is resistance? If you attach wires from the two terminals of a battery (or the two prongs of an electrical outlet plug) to the ends of a metal bar, an electrical current (flow of charge) moves through the bar. The battery or outlet (known technically as a voltage source) acts to drive the current through the metal. The amount of current that flows depends on that property of the metal called its resistance (you might think of resistance as the tendency to impede a current flow). Metals typically have low values of resistance (except very thin wires), whereas things like glass have high resistances.

Kamerlingh Onnes asked the question: What happens to the resistance of a metal as its temperature becomes very low? He already knew that a metal's resistance generally decreases as the metal gets colder. Would this continue until absolute zero, or would the decrease level off? Would the resistance disappear at absolute zero, or would some residual resistance remain? Experiments with platinum showed a leveling off, but this seemed to be caused by impurities. So they tried mercury next, because mercury could be highly purified by distillation. The behavior of mercury's resistance was totally unexpected. The resistance slowly decreased as usual until a temperature was reached just below that of liquid helium. Then, the resistance dropped abruptly to zero! This was so extraordinary that the results were initially misinterpreted as a mistake, a short circuit in the apparatus. As luck would have it, a technician monitoring the temperature dozed off and the temperature drifted upward through the transition point, so the experimenter actually witnessed the abruptness of the change. Subsequent careful experimentation revealed that the transition to zero resistance really did occur as a sharp drop at a specific temperature, now called the critical temperature. Superconductivity had been discovered.

The initial discovery occurred in 1911. Over the next few years, superconductivity was found in several other metals and explored in more detail (for example, very high currents and magnetic fields were found to destroy the effect). In order to determine whether the resistance was really zero or just very small, currents were set going in superconducting rings and left alone; these currents persisted indefinitely, showing that the resistance was truly zero. Early efforts were made to understand superconductivity on a theoretical basis, but these efforts were only partially successful, yielding important insights but no comprehensive theory. After nearly half a century, John Bardeen and his colleagues finally devised a genuinely coherent explanation for superconductivity that accounted for all of the experimental findings.

§2. THE MICROSCOPE

In scientific usage, we don't always use the word "observation" to mean something you actually see with your eyes. But it's still true that sight is one of our primary means of acquiring information; and in some cases, an "observation" really *is* an observation. If we are indeed using our eyes as scientific instruments, then we'll want to improve them. The primary tool for this purpose is the lens. A lens is a curved piece of glass that bends rays of light passing through it (by refraction at the curved glass surface). The action of lenses was known to Arab science a millennium ago. In Europe, Roger Bacon wrote about using lenses to magnify letters as an aid to reading, and eyeglasses were in use by 1300. The art of lens grinding continued to improve, and both telescopes and microscopes existed in primitive forms before 1600. Our eyes had been improved for seeing things that are far away and for seeing things that are very small. Our power to make scientific observations was thus increased, and astronomers (led by Galileo; see chapter 5) quickly exploited the telescope in their work. At about the same time, a few pioneers began to look with microscopes at small things like insects, spores, and fabric threads in order to see more detail.

Many of the early microscopes were limited in their magnifying power by the difficulty of grinding the small highly curved lenses needed. Neither the compound microscopes with more than one lens (similar to modern versions) nor the simple single lens microscopes had magnifications much greater than about 50X. Nevertheless, a lot of interesting work was done with such instruments, the most famous example being Robert Hooke's series of studies published as *Micrographia* in 1665. The objects of study in all this work were visible to the unaided eye, but only as small blurs (fleas, for example). The microscope gave us a wealth of new and detailed information about such things. But the really amazing discovery, which is the main topic of this section, was yet to come, awaiting the ability to craft a better lens.

Improved lenses were soon produced, and some of the best were ground by the premier lensmaker of that century, Antony van Leeuwenhoek. We still don't know how he produced such fine lenses, for he never divulged his secrets to anyone. Leeuwenhoek's microscopes were relatively simple, consisting of a single lens embedded in a small piece of metal with a screw to position the sample at the proper position for focus. The simplicity of the design was paid for by the difficulty of using such microscopes in practice. Leeuwenhoek devised new techniques, requiring extreme patience and skill, for using his handmade microscopes (he also refused to share these techniques with anyone else). The results of all this work and skill were remarkable: microscopes with magnifications of over 300X and an extensive series of papers describing his observations spanning 50 years. In 1674, Leeuwenhoek looked at some scummy pond water with one of his high-power microscopes and thereby revealed a completely new world, unknown and unexplored. He saw through his lens a swarm of tiny creatures with strange and wonderful forms. They had a variety of sizes and shapes, some were translucent and some glittered, they moved slowly or quickly using rows of beating cilia or spiral-shaped appendages. Within this droplet of ordinary water, an entire unsuspected universe of life was going about its business in the microworld, which could now for the first time be seen.

Leeuwenhoek's discovery of the microbes in pond water was one of the high points in his illustrious career. During this career, he also discovered bacteria, spermatozoa, and blood capillaries; and he studied in detail the cell structures of plants, red blood cells, embryos, muscle fibers, and insect anatomy. He discovered a whole new world and spent his life carefully observing and documenting the facts of this world. Leeuwenhoek and his fellow microscopists were far ahead of their time; it took the science of biology about a century to formulate a systematic understanding of nature into which the microbial world could fit. Their techniques and observations were the beginnings of what we now call microbiology, and we now know that the denizens of the microbial world are very important to us (for both good and ill). Although we now take the existence of microbes for granted (what five-year-old doesn't know about germs?), it's worth remembering that no one had ever seen a microbe before Leeuwenhoek. Because no one had seen microbes, they essentially did not exist (in a conceptual sense). The key that allowed Leeuwenhoek to enter this new world was his improvement to the art of lens grinding and the increase in magnification this gave his microscopes.

§3. Radio Astronomy

While the microscopists used lenses to see the very small, the astronomers used lenses and mirrors to see farther into the night skies. Making bigger and better telescopes, astronomers discovered new planets, moons around other planets, galaxies, and stars from ever farther away. But all these discoveries are not the story I'm going to tell right now, because even the biggest and best lenses and mirrors still gather ordinary light, the same light seen by our unaided eyes. Our present story is about looking at the skies with a completely different instrument, and seeing a world that is invisible to our eyes at any magnification.

The story starts in 1930. The first transatlantic telephone link using radio waves had recently been constructed, and the sound quality left a lot to be desired. Bell Telephone Laboratories assigned Karl Jansky to track down the cause of all the static and hissing sounds in the lines. He found that most of it was caused by thunderstorms, but a small residual hiss seemed to be coming from outer space. Jansky eventually localized the source of these radio emissions as the Milky Way, but he was unable to pursue these studies any further (Bell assigned him to other duties).

In retrospect, this was the origin of radio astronomy. But the field did not take off very quickly. Astronomers were totally uninterested, and the only person who continued Jansky's work (a radio engineer named Grote Reber) did so as a hobby. Astronomers paid little attention for at least three reasons: the cause of the radio emissions was completely mysterious; astronomers had no expertise in radio engineering (in contrast to telescope construction, which was a highly advanced specialty by this time); and the source of the radio waves could not be localized very precisely. This last problem was a very serious one. Traditional astronomy had progressed to a remarkable degree of precision (comparable to locating the position of a dime over a mile away). The early attempts at radio astronomy couldn't really locate anything; they merely indicated the general direction from which the radio signals came. Astronomers were unimpressed by such large bloblike indications of position. But work on radar during World War II greatly improved radio technology, and the precision of the interstellar radio source locations continued to increase (mostly by making the radio telescopes, which were basically antennas, much bigger or by combining the signals from widely separated antennas). In the postwar period, the radio astronomers began to locate radio emission sources with positional accuracies good enough to attract the attention of the traditional astronomers, and the field began to make serious contributions to science. At this point, we had a revolutionary new tool with which to look at the universe, and the radio universe looked a lot different from the optical universe we had seen for thousands of years. Galaxies that had appeared to be shining calmly turned out to have enormous radio sources at their cores, shooting out violent jets of material across millions of light-years. The optically dim remnants of a supernova shine brightly with radio waves, as do other forms of star death and birth. A number of faint stars, called pulsars, emit strong radio waves with intensities that vary periodically with clockwork stability. All of these things cannot be seen with any eye or telescope. They are part of a world that we can't see, a world whose existence we never suspected.

As the techniques of radio astronomy matured, its results had to be taken seriously. But these results still remained mysterious and difficult to interpret. A particularly bizarre result was discovered in 1960 by radio and optical astronomers working together: the discovery of quasars. The initial discovery was a brightly emitting radio source that was extraordinarily small in size. The high resolution of the new radio telescopes allowed radio atronomers to see how small the source was and also to locate it precisely in the sky. Knowing an exact position allowed the optical astronomers to train their telescopes on that spot in order to see what it looked like in the world of light. Remarkably, the object just looked like an ordinary star. Worse yet, the star appeared to be 4.5 billion light-years away. (The method used to determine this distance is a little too involved to relate here; it involves the spectroscopy of starlight, Doppler shifts of frequency, and the expansion of the universe.) But for something that bright to be so far away, it must throw off more energy than 100 galaxies put together. Yet, the size of the object was only a few hundred times bigger than our solar system. The strange new objects that had been discovered were called quasi-stellar radio sources, which was shortened to quasars. How can we understand the properties of these quasars? How can something that small radiate that much energy? Just as the radio and optical astronomers had teamed up to discover quasars, the astrophysical theorists teamed up with the general relativity experts to understand what quasars are. We now believe that quasars are supermassive black holes, equivalent to millions of stars. (A black hole is an object that has suffered "gravitational collapse"; its gravitational field is so strong that not even light can escape, thus the name.) As matter is sucked into the giant black hole, the matter accelerates and radiates away the energy we see. Although theorists are still investigating these processes, the consensus is that this explanation is substantially correct, and that a similar picture explains the observations of radio galaxies. The huge gas jets streaming from the cores of radio galaxies are caused by spinning black holes that are somewhat less massive than the quasar black holes.

For thousands of years, astronomy was done with our eyes. Telescopes extended the powers of our eyes to see farther, but radio astronomy extended the powers of our senses to see a new and invisible universe. In this new universe, we discovered things like quasars, which we hadn't even imagined before. Karl Jansky's discovery of radio waves from the stars opened up a new world to explore, and moved black holes from the dustbins of theorists into the hearts of galaxies.

FOR FURTHER READING

Microbe Hunters, by Paul de Kruif, Washington Square Press, 1965 (originally Harcourt, Brace, 1926).
Single Lens, by Brian J. Ford, Harper & Row, 1985.
The Invisible Universe Revealed, by G. L. Verschuur, Springer-Verlag, 1987.
Twentieth Century Physics, edited by Laurie M. Brown, Abraham Pais, and Sir Brian Pippard, Institute of Physics Publishing and American Institute of Physics Press, 1995.
"The Discovery of Superconductivity," by Jacobus de Nobel, *Physics Today*, vol. 49, no. 9, p. 40, September 1996.

EPIGRAPH REFERENCE: P. W. Bridgman, *The Way Things Are*, Harvard University Press, 1959, p. 149.

Chapter 4

CLOSE, BUT NO CIGAR: DISCREPANCIES
AS A TRIGGER TO DISCOVERY

> Indeed, I have seen some indications that the anomalous
> properties of argon are brought as a kind of accusation
> against us. But we had the very best intentions in the
> matter. The facts were too much for us, and all that we
> can do now is apologize for ourselves and for the gas.
> *(Lord Rayleigh)*

S EEING what you do not expect can be a powerful impetus to develop new ideas. Sometimes this process is obvious. If you believe that yellow fruits don't exist, then seeing a banana will modify your beliefs. This is progress; it's a primitive example of scientific discovery resulting from an observed discrepancy. Most examples are more complicated. There are a number of episodes in the history of science where initially small discrepancies eventually instigated major scientific revolutions. Two celebrated cases in physics are the Michaelson-Morley experiment (which was eventually explained by the theory of relativity) and the blackbody radiation measurements (which were the beginning of quantum mechanics). I've chosen not to discuss these two cases here because they are rather complicated and because so much has been written about them elsewhere. Instead, we'll look at some simpler examples. In the first example (argon), a tiny numerical discrepancy in a routine measurement provides a clue that leads to the discovery of a new gas, and this gas turns out to be a member of a previously unsuspected class of chemical elements. In the second example (the barometer), a somewhat minor but unexplainable observational discrepancy leads to an important new concept (atmospheric pressure). In the last example (Neptune), a remarkably small discrepancy leads to the vindication of a theory (classical mechanics) and the discovery of a new planet.

§1. ARGON

Toward the end of the nineteenth century, the composition of the atmosphere was well known to be oxygen and nitrogen, plus small amounts of carbon dioxide and water vapor. Around 1890, the famous British physicist Lord Rayleigh was making some measurements of the densities of oxygen and nitrogen. These measurements were intended to increase the precision with which the densities of the gases were known. Rayleigh didn't anticipate anything radically new to result from these experiments. But his density measurements for nitrogen did result in a small and puzzling discrepancy. He made one of his nitrogen samples by starting with air and removing the oxygen (and other reactive gases). He measured its density. He made another nitrogen sample by starting with ammonia and decomposing it. For this sample, he measured a different density (slightly smaller). The difference between these two density measurements was not large, only about one part in a thousand. But because he had measured the densities with such high precision and repeated the experiments many times, he knew the difference was real even though it was small.

At first, Rayleigh assumed that there must be something different about the nitrogen in the two cases. One suggestion, for example, was that some of the heavier nitrogen exists as a triatomic molecule instead of the normal diatomic nitrogen molecule (just as ozone is a triatomic oxygen molecule). But this idea didn't really stand up to scrutiny because when he tried to make the heavier nitrogen by any other method, he failed. Only nitrogen in the atmosphere had an anomalously high density. Eventually, Rayleigh came up with the idea of chemically reacting the atmospheric nitrogen in order to remove it (as a gas) from the container it was in. He discovered that there was a small residue of gas that he couldn't remove. A chemist named William Ramsay also started to experiment with removing nitrogen from atmospheric nitrogen samples, and he obtained similar results, that is, a small amount of some gas that couldn't be removed.

Rayleigh and Ramsay came to the conclusion that they had discovered a new and previously unsuspected gas in the atmosphere. This gas had some remarkable properties: The density of the gas was much higher (i.e., the gas was much heavier) than any other gas known at that time. In addition, the new gas was totally unreactive chemically; it was unaffected by strong acids, strong bases, and highly reactive metals. They named their new gas "argon." The scientific community was astonished by this discovery, and initially quite skeptical. But the detailed research of Rayleigh and Ramsay (including measurements of argon's density, index of refraction, solubility, and atomic spectra) built a convincing case, and the scientific community soon hailed the discovery of argon as an exciting event. Argon was the first of the so-called inert gases to be discovered. Ramsay went on to find a number of others (krypton, neon, xenon, and helium), and a new column of the periodic table (see chapter 2) was established for this new class of gaseous elements that didn't form chemical compounds. Thus ended the story that started with a minor discrepancy in the density of nitrogen.

There are some interesting historical footnotes to this story. Helium had actually been "discovered" decades earlier (1868), in the sense that its atomic spectra had been measured in the light from the sun. Since these atomic spectral lines didn't belong to any known element on earth, the scientists who observed them (Lockyer and Janssen) interpreted the spectra as belonging to a new element found in the sun. Lockyer named the new element helium after the name of the Greek sun god. Many years later, after argon was discovered in 1894, Ramsay found helium on earth (recognized by its atomic spectra) in the course of his search for inert gases. The other interesting bit of history is that these gases are not actually completely inert; in the 1960s, chemists were finally able to create a few exotic chemical compounds containing these gases, which until then had been thought to be totally unreactive.

§2. THE BAROMETER

Our next example is also about air, but this example is from a much earlier time, namely 1643. Oxygen and nitrogen haven't been discovered yet and won't be for over a century. Very little, in fact, was understood about air at that time. One important property of air was known to advanced thinkers, though it was still debated by reactionary academics: it was known that air had weight. The fact that air had weight was known in particular by Galileo's talented student, Evangelista Torricelli. Torricelli (and his colleague Viviani) became interested in a somewhat peculiar minor discrepancy between an observed fact and the accepted explanation for that fact. The fact that intrigued them was that water could only be pumped to a height of 34 feet by suction pumps. This was well known to plumbers, mining engineers, and other practical folks who had to move water from lower to higher places. These suction pumps were like the old hand pumps with which rural Americans got the water from their wells. Pumps like these operate on the same principle you use when you suck a liquid up a straw. In other words, if you keep making your straw longer and longer, you won't be able to suck water to the top once your straw is over 34 feet long. Why should this be true?

In order to answer this question, you need to answer a prior question: Why does the suction make the liquid rise at all? At that time, the accepted explanation for why suction makes a liquid rise was the one given by Aristotle. The explanation was based on a very general principle, namely that nature abhors a vacuum (*horror vacui* in Latin). In other words, an empty space is an unnatural state, and nature will therefore attempt to fill this space with something. In the case of the pump, applying suction starts to create a vacuum, and so the liquid rises to fill the space where the vacuum started to form. (Just to avoid any confusion, I should say right away that all this is now considered nonsense. The phrase "nature abhors a vacuum" still survives as a literary device, but it has no scientific content. Most of the universe is a vacuum.) So, it was thought that a suction pump worked because nature abhorred a vacuum. But the suction could only pump water up to 34 feet. This height limitation is the discrepancy between the observed fact and the explanation! Why should nature's abhorrence of a vacuum only extend up to 34 feet?

Torricelli pondered over this discrepancy, looking for a way to make sense of it. We have no record of his thought processes, but somehow he made a connection between this discrepancy in the theory of suction pumps and the fact that air has weight. Making this connection was a key insight, and working out the consequences of his idea allowed Torricelli to discover atmospheric pressure, to invent the barometer, and to create a vacuum (all in the same experiment!). Let's take a look at his reasoning. If air has weight, and if we are living at the bottom of what might be called an ocean of air, then the weight of this large amount of air above us must be pressing down on us. (We now call this pressing down by the name "atmospheric pressure," a concept that did not exist until Torricelli thought of it.) The weight of the air also presses down on the surface of the water to be pumped. If we now remove some of the air over the water in a tube (by suction), the weight (or atmospheric pressure, as we'd say now) is less over the water in the tube. The pressure of the air over the rest of the water then pushes the water up the tube, since the pressure in the tube is less. Note that as the water column rises in the tube, the weight of the water column itself pushes back on the pressure that is pushing the column up. The higher the column, the more it weighs and the harder it pushes back. When the weight of the water column is pushing down just as much as the atmospheric pressure is pushing it up, then the column won't rise any higher. This happens when the column is 34 feet high. Torricelli's explanation for the observed discrepancy was not merely fine-tuning of the currently accepted thinking. He proposed a completely new conceptual model for the action of the pump. The idea of atmospheric pressure had never been thought of before because the effects of atmospheric pressure are not usually apparent. Replacing abhorrence of a vacuum with atmospheric pressure was, in its own way, a scientific revolution.

If we are surrounded by a strong pressure due to the air, why don't we notice it? We don't feel like we're being pressed on. We don't typically see or feel any effects of this pressure. The reason atmospheric pressure has so little visible effect is that the air is both inside and outside most objects, including ourselves; the inside and outside pressures cancel, and therefore the net pressure is zero. If you want to see a dramatic demonstration of the effects of atmospheric pressure, try this: Put a little water into an empty aluminum can and heat the water to boiling for a few minutes (this will fill the can with water vapor). Then quickly plunge the can, open end downward, into cold water. The coldness will condense the water vapor into liquid, leaving a vacuum inside the can. Atmospheric pressure will then rapidly implode the can with a loud bang, leaving a crumpled bit of metal.

But none of this information was known to Torricelli and his contemporaries in 1643. Torricelli's radical new idea was certainly interesting, and it did explain the limitations of suction pumps; but his idea was, so far, just unproven speculation. What separates vaguely interesting speculation from true science is the next step. Torricelli reasoned that if the weight of the water was responsible for the height limitation of the water column, then a new liquid with a different density (i.e., different weight) should rise to a different height. In fact, the ratio of the densities should predict the height of the new liquid column (e.g., a liquid that is twice as dense as water should rise to only half the height of the water column because the weights of these two columns are the same). So Torricelli did an experiment. He knew that the density of mercury is 13.6 times greater than the density of water. If his explanation was correct, atmospheric pressure should only be able to lift a column of mercury up to a height of 2.5 feet (34/13.6 = 2.5). A diagram of the experiment is shown in Figure 2. The procedure is to take a glass tube that is closed at one end and fill the tube with mercury. The open end of the tube is then placed in a bowl of mercury, with the tube held vertical and the closed end at the top. The column of mercury then drops under its own weight, leaving a vacuum in the space of the tube above the mercury. The mercury column drops until it reaches the height at which its weight is counterbalanced by the pressure of the atmosphere on the bowl of mercury. The height of the mercury column turned out to be 2.5 feet, just as Torricelli had predicted. A vacuum, of which nature was supposed to have such an abhorrence, had been easily created at the top of the tube. The concept of atmospheric pressure was strikingly confirmed.

Two important instrumental techniques were invented at the same time in this experiment. Now that it was possible to create a vacuum, it was also possible to do experiments in a vacuum. This capability led to entire new experimental programs (although the preferred method of making a vacuum became, after a few decades, the newly invented air pumps). The second important technique derives from the following fact: the height of the mercury column is directly proportional to the atmospheric pressure exerted on the bowl (see chapter 19). Measure the height and you've also measured the pressure. Thus, the inverted tube of mercury used in this

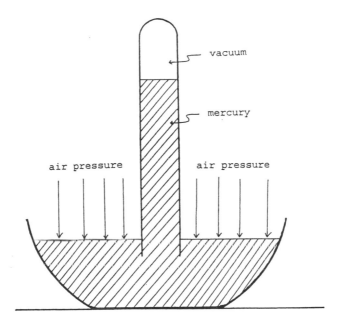

Figure 2. A schematic version of Torricelli's experiment, testing his prediction that atmospheric pressure could only keep a column of mercury 2.5 feet high. In performing the experiment, he also invented the barometer, since this device can also measure changes in atmospheric pressure.

experiment is, automatically, a pressure-measuring instrument. We call this instrument a barometer. The mercury barometer, in essentially the same form as Figure 2, is still occasionally used to measure variations in atmospheric pressure. Why are there any variations in atmospheric pressure? One reason is that the pressure varies with altitude. Since there is less air (less weight) above you when you are higher up, a pressure decrease with altitude is predicted by Torricelli's theory. In fact, a celebrated and important test of the theory was carried out in 1648 by Pascal, who cajoled his brother-in-law into hauling a barometer up to the top of a mountain. This experiment demonstrated that the atmospheric pressure decreased at higher altitudes as predicted. Variations in atmospheric pressure are also associated with changes in the weather. Low pressure is typical of stormy weather and high pressure is typical of cool clear weather. For this reason, televised weather reports usually tell you the current atmospheric pressure and how it's changing. You may have noticed that the atmospheric pressure is reported in inches. This refers to the height a mercury column in a barometer would reach, using an inch of mercury as a unit of pressure.

Having instruments to measure gas pressure and the ability to create a vacuum were invaluable capabilities for science. These new experimental capabilities, in addition to the important new concept of the atmosphere as a sea of air that exerts pressure, led to many productive avenues of research and a great deal of scientific progress. All that progress began with a minor observational discrepancy with a well-accepted but incorrect principle.

§3. NEPTUNE

Our last example is the discovery of the planet Neptune in 1846. We discuss the motions of the planets extensively in chapter 5, ending with Newton's explanation of these motions. Between the time of Newton's work and the discovery of Neptune, the intellectual landscape had changed profoundly, as had the celestial landscape. During Newton's lifetime, his work was an exciting beginning for a new science. By the mid-nineteenth century, Newton's mechanics was a highly developed and sophisticated science, and also an established orthodoxy. Meanwhile, in the sky, the five planets of the ancient Babylonians had been joined by an exciting newcomer in 1781: the planet Uranus. The discovery of Uranus by William Herschel isn't our topic at the moment, but it does present a number of interesting features. In some ways, it was a discovery by serendipity, since Herschel just happened by accident to be looking at the spot where Uranus was. But it's also an example of discovery resulting from better instrumentation because he was only able to see that Uranus is a planet (instead of a star) as a result of using a telescope with very high magnification. Even with such high magnification, a talented and experienced observer like Herschel was required in order to see the subtle difference.

Why am I discussing Uranus when our topic is Neptune? Uranus actually plays a key role in our story. After Uranus was discovered, its orbital motion was recorded over many years (orbital information was also found in historical records, where it turned out that Uranus had been seen before but mistakenly identified as a star). Newton's laws of motion were used, as usual, to calculate a predicted orbit for Uranus. Between the observed orbit and the calculated orbit, however, there was a very tiny discrepancy. Now the calculation of planetary orbits had become a highly refined and exact science by the middle of the nineteenth century. The earliest applications of Newton's laws to planetary motion assumed that the planets were influenced only by the gravitational attraction of the sun. Because the sun is many hundreds of times more massive than any planet, only considering the sun is a very good approximation. But all the planets have some mass, and so they all have some slight gravitational influence on each other. As long as the observations of the planetary positions are no more precise than the predictions of the approximate theory, these slight effects can be ignored. As the power and precision of telescopes improved, however, astronomers were able to measure the positions of the planets so precisely that the theory could no longer make accurate predictions using only the gravitational influence of the sun. The other planets had to be included.

Including the gravitational influence of the other planets is extremely difficult. Applying Newton's laws to calculate the motion of objects can only be done for (at most) two mutually interacting objects. The reason for this limitation is mathematical; the equations we get for three or more objects can't be solved. Fortunately, the great mass of the sun makes it initially possible to ignore the other planets, and only that fact made it possible for Newton to apply his theory. As I said, subsequent progress in observational astronomy eventually forced physicists to deal with this problem, which they did in a remarkably clever manner. They couldn't solve the "three body problem" directly, but since the gravitational attraction of a planet is so much smaller than that of the sun, they were able to treat it as a small *correction* on the orbit they *could* solve using only the sun. The technical jargon for this technique is that the gravity of the planet is treated as a small perturbation on the gravity of the sun. The effect of this perturbation on the orbit could be approximately calculated using special mathematical methods that were worked out for this purpose. A feature of the technique is that the approximation becomes better as the perturbation becomes smaller (see chapter 6 for a more general discussion of this idea). For this reason, the technique works very well for our planetary motion problem. Needless to say, the actual mathematical manipulations are exceedingly difficult. Even so, a number of brilliant mathematical physicists (Laplace being the most prominent) had used these techniques to work out all the motions of the planets by the beginning of the nineteenth century, and all of the predictions fit the observations perfectly.

Until Uranus. By the 1840s, it had become clear that there was a small discrepancy between the growing collection of observations and the increasingly precise perturbation calculations for Uranus. When I say small, keep in mind just how good the observations and calculations had become by that time. The discrepancy we are discussing is roughly 30 seconds of arc, about the same as the angle between two automobile headlights which you see from five miles away. But as small as the discrepancy was, it was still real and it shouldn't have been there. How could this discrepancy be explained? One "obvious" explanation is that the inverse-distance-squared relation for gravity breaks down in the case of Uranus and some different law holds (perhaps due to its much greater distance from the sun). Such a revolutionary and arbitrary conclusion is unwarranted by the facts of this case, however; throwing out universal gravitation should be our last resort, not our first choice (see chapter 12). Another explanation might be that a large body (e.g., a comet) had passed close to Uranus sometime in the previous hundred years, greatly perturbing its orbit. This hypothesis is plausible, but it has the disadvantage of being untestable (see chapter 14). Other possible explanations might be invented, but the most popular proposed explanation at the time is also the one we are most interested in here: Another planet exists in our solar system, exerting a gravitational perturbation that accounts for the discrepancies in the orbit of Uranus.

If you make the hypothesis that another planet, yet unknown, is the cause of the irregularities in the orbit of Uranus, then your next step would naturally be an attempt to locate the new planet. In particular, you would want to calculate the orbit that this new planet must have in order to account, quantitatively, for these irregularities. The difficulty of such a calculation is mind-boggling. It's hard enough to calculate the perturbations on an orbit from a source that you know, but here you must work backwards and figure out the location of the source based on the perturbations. Yet two formidable mathematicians, J. C. Adams in England and U. LeVerrier in France, undertook this calculation and they both came to the same conclusion. The world now had a prediction of where to look for the hypothetical new planet. LeVerrier sent his work to the observational astronomer J. G. Galle in 1846. Galle immediately set to work looking for the planet. He found it on his first night of observation! The new planet eventually came to be named Neptune. This exciting discovery was a great triumph for Newton's mechanics.

Some interesting peculiarities can be found in this historical episode. Adams actually completed his calculations before LeVerrier and sent the results to G. Airy, the Astronomer Royal. Airy inexplicably paid little attention to this work, and thus missed the chance to discover Neptune. Another interesting point is that both Adams and LeVerrier made some very incorrect assumptions about Neptune's orbit at the beginning of their calculations. The orbits they calculated were therefore not accurate. Because they only possessed recorded orbital positions for a limited period of time, however, the orbit of their calculated planet and the orbit of the real Neptune were not yet very far apart. This circumstance, in combination with a dose of good fortune, made their predicted position for Neptune very close to its actual position in 1846. These historical curiosities need not distract us from the main point. A remarkably small discrepancy in the orbit of Uranus led directly to the discovery of a new planet, Neptune. Along the way, Newton's laws of motion and universal gravitation were verified to an unprecedented degree of precision.

FOR FURTHER READING

Astronomical Discovery, by H. H. Turner, University of California Press, 1963 (originally Edward Arnold, 1904).

On Understanding Science, by James B. Conant, Yale University Press, 1947.

Springs of Scientific Creativity, edited by R. Aris, H. T. Davis, and R. H. Stuewer, University of Minnesota Press, 1983.

Serendipity: Accidental Discoveries in Science, by R. M. Roberts, John Wiley & Sons, 1989.

EPIGRAPH REFERENCE: Lord Rayleigh, quoted by John N. Howard in *Springs of Scientific Creativity*, edited by R. Aris, H. T. Davis, and R. H. Stuewer, University of Minnesota Press, 1983, p. 178.

Chapter 5

INGREDIENTS FOR A REVOLUTION: THEMATIC
IMAGINATION, PRECISE MEASUREMENTS, AND
THE MOTIONS OF THE PLANETS

> We see [Kepler] on his pioneering trek, probing for
> the firm ground on which our science could later build,
> and often led into regions which we now know to be
> unsuitable marshland.
> *(Gerald Holton)*

THE DISCOVERY of how and why the planets move, related in this chapter, is intimately tied to the beginnings of modern science. It's worth remembering, however, that modern science didn't exist until after this story had already ended. For that reason, the motivations and thinking patterns of the people who contributed to our understanding of the motions of the planets were vastly different from our modern-day point of view. An understanding of the planetary motions emerged only slowly from a strange tangle of metaphysical assumptions, thematic hypotheses, and observations of the night sky. When a true understanding was finally achieved, physics as we know it came into existence as part of the same event. The story starts with the ancient Babylonians and ends over four thousand years later with Sir Isaac Newton. The Babylonians were careful observers of the night sky, recording the positions of the moon, stars, and planets. They also developed sophisticated mathematical techniques to help them find the regularities in the paths of the celestial lights. But their understanding of these lights was not scientific in the way we understand the word; the night sky of the Babylonians was a place of myth and wonder, populated by living beings. Other civilizations had also developed such an observational, mathematical, and mythological astronomy to a high degree (the Mayans and the builders of Stonehenge come to mind, for example). The Babylonians hold a special interest for us, though, because their legacy of observational and mathematical work was passed on to the Greeks.

§1. GREEK ASTRONOMY

The Greek philosophers had inquiring minds; they wondered about the nature of these lights in the sky and questioned their ancient religious beliefs. Instead of merely using mathematics as a practical tool to predict where the moon or a planet would be on a certain day, the Greek mind also wanted to know the underlying causes. The idea that the true nature of the world is based on mathematical relationships originated with Pythagoras of Samos, one of the greatest thinkers of antiquity, who lived in the sixth century B.C. Pythagoras and his followers combined mathematics, number mysticism, and science, applying their ideas to music, astronomy, and medicine. Astronomers who were followers of Pythagoras continued for several generations to create world systems in order to explain the motions of celestial bodies in a way that was consistent with their philosophy. The last of this line of thinkers was Aristarchus; the system he proposed was that the earth and the planets move around the sun, with the earth rotating on its axis. In other words, Aristarchus created a system that was conceptually identical to the system Copernicus devised eighteen hundred years later. Unfortunately, his ideas did not catch on. No one really knows why this happened, but Greek astronomy instead became dominated by the earth-centered systems that we'll look at next.

The Pythagorean emphasis on the importance of mathematics was incorporated into the thinking of Plato. The immense influence of Plato on European culture kept this mathematical ideal alive, and it eventually was incorporated into modern science. But Plato's philosophy is based upon idealizations rather than the study of nature, upon pure thought rather than experience and perception. An example of this idealized thinking that is relevant to our story is Plato's teaching that the sphere and the circle are perfect shapes in some metaphysical sense. Many astronomical observations indicate that celestial bodies move in circular paths. Also, the circle is a mathematically simple figure, making it highly useful in the application of geometry to astronomy. These two facts, in combination with the philosophical idea of the circle as a perfect shape, resulted in an almost inescapable conclusion: The planets must, by all of this reasoning, move in perfect circular paths. But there is a problem with this conclusion. If the planets are moving in perfect circles around the earth, we should always observe them moving across the sky in the same direction at a constant rate. The planets don't move this way, however. They occasionally stop, move backwards for a while, and then return to their usual direction. This so-called retrograde motion presents a problem to the believer in circles. The problem was solved in an ingenious manner by Eudoxos, a student of Plato, using sets of nested spheres. This kind of think-ing eventually culminated in the famous system of Ptolemy: the planets move in little circles called epicycles, while the epicycles themselves move on larger circles around the earth. This system results in a path for the planet that allows retrograde motion. Ptolemy could then account for all the observed motions of the planets using only the circular paths allowed by philosophy. The sun, moon, and five planets could all be described using a total of thirty-nine circles. The geocentric system of Ptolemy, with all of its wheels-within-wheels, was the last major innovation in astron-omy for fourteen centuries. As you undoubtedly know, the civilization of antiquity fell into decay at about that time, while Western Europe entered into the Dark Ages for hundreds of years. The learning of the Greeks was preserved in the East and was later assimilated by the civilization of Islam. Through contact with the Islamic empire, Europe rediscovered the teach-ings of Ptolemy and Aristotle in the twelfth century.

The system of Ptolemy can be thought of as just a mathematical device for tracking the planets' motions, but in fact it was not thought of this way. Instead, Ptolemy's work was combined with the physical ideas of Aristotle, in which the celestial bodies are attached to crystal spheres whose motion carries the planets and stars along their paths. The motions of the spheres were considered natural and thus did not need further ex-planation. The earth, at the center of the cosmos, was governed by differ-ent laws; on earth the natural motion was to be at rest (or, if elevated, to fall down). Also, on earth there was change and decay whereas the celes-tial regions were immutable and perfect. Outside the outermost sphere of stars was the Prime Mover who kept the whole thing going. The science of Aristotle is self-contained and logical, in its own way. It is completely nonmathematical, though, and the mathematical astronomy of Ptolemy is simply grafted on. It may seem that we have wandered off the subject a little with all these digressions about Greek philosophy. As we proceed, however, you'll see that such philosophical considerations play a central role in the story. The teachings of Aristotle were incorporated into Chris-tian theology, and they dominated European thought for about three hun-dred years. But eventually the closed and stable world of the High Middle Ages began to break up in the brisk intellectual winds of the Renaissance.

§2. COPERNICUS

One of the ironies of history is that a great revolution in human thought, the banishing of the earth from the center of the universe, was instigated by a quiet and conservative thinker who actually disliked novelties. Most of the thinking of Nicolaus Copernicus was quite medieval in character, and he saw his work in astronomy as part of the humanist tradition in which he and his contemporaries were engaged. This humanist activity consisted of recovering and translating the writings of ancient Greece. The humanists were making available, along with literature and philosophy, the works of the Pythagoreans, of Archimedes, and other scientific or mathematical treatises. Copernicus seems to have viewed his work as a restoration of the ancient Pythagorean astronomy rather than as a radical new idea. Copernicus was educated in Italy, a center of advanced Renaissance thinking. The idea of a sun-centered universe was known and discussed there. Copernicus did not invent the heliocentric system. His great accomplishment was to work out the mathematics of the heliocentric system as methodically as Ptolemy had done for the geocentric system. This methodical working-out was a necessary step before astronomers would take the heliocentric idea seriously.

Another indication of Copernicus' conservatism is that he used, with few exceptions, only the observations of the Greeks recorded in Ptolemy's work (which in fact were not all that accurate). Contrast this behavior with the program of Regiomontanus, a prominent astronomer who lived a generation before Copernicus. Regiomontanus believed that Ptolemaic astronomy was flawed, and he also considered the merits of the heliocentric system; but instead of reworking the mathematics of Ptolemy, he chose to build better observational equipment to make new and better measurements of the positions of the planets. In a sense, Regiomontanus was more modern in his outlook than Copernicus, but it was Copernicus who started the revolution in astronomy. On the other hand, we'll see later that improved measurements were also needed to advance the cause.

In thinking about Copernicanism, we need to keep in mind that three different levels of meaning are involved. First, there is the mathematical level, the details that are primarily of interest to astronomers; second, there is the physical level, the assertion that the earth actually moves around the sun and rotates on its axis; third, there is the cosmological level, the implications of destroying the closed and stable world of Aristotle where everything has a natural place (and remember, this was the world in which medieval folks lived). Let's discuss each of these levels in more detail.

The mathematical details of his system were clearly the major interest of Copernicus himself. As a method for making astronomical calculations, the new system made a few improvements and simplifications. Professional astronomers used Copernicus' work to make new tables of planetary positions, and it was helpful for practical problems like navigation and calendar reform. But this kind of calculational simplification is not really the main point. A more important issue is conceptual simplicity, and here Copernicus' system has both good and bad points. The major simplification provided by heliocentrism is that it easily explains retrograde motion. Retrograde motion is just an artifact caused by the motion of the earth, a kind of optical illusion. For example, during one year the earth travels all the way around the sun (thus going in two opposite directions), while a planet like Jupiter only goes through a small part of its orbit (always in the same direction). So, looked at from the earth, Jupiter *appears* to change direction but it really doesn't (see Figure 3). Other observations are also explained in a satisfying manner, such as the fact that Mercury and Venus are always near the sun. But there are also some problems. Because the planets don't actually move in circles (more on this later) and because some of the old Greek observations were mistaken, Copernicus had to reintroduce epicycles so that his system would match the positions of the planets. In fact, Copernicus used about the same number of epicycles as Ptolemy. Also, the center of the system was not actually the sun but instead was a point in empty space near the sun.

These subtle points were mostly the concern of astronomers and mathematicians, not the typical educated person (humanists, clergy, academics, aristocrats, etc.). For most people, the question of whether the earth moves was a more interesting and compelling question. The debate was lively, but not widespread. To most people in the generation following Copernicus, the earth's motion did not really seem to be a pressing issue. The reaction among Protestants was mixed, and the Catholic Church was fairly favorable (Copernicus was a Canon of the Church), though neither group paid much attention at first. Among freethinkers, innovators, and intellectual radicals of various sorts, the Copernican idea of a moving earth was widely embraced simply because it was so contrary to tradition. The dedicated Aristotelian scholars, on the other hand, were vehemently opposed to the idea; heliocentrism contradicted Aristotle's teachings, violated his cosmology and his physics, and seemed patently absurd. (A stupid idea. Do you feel like you're moving?) Many arguments, based on Aristotle's physics, were put forth against a moving earth. Copernicus believed in the motion of the earth based on his mathematical astronomy, even though his physics was purely Aristotelian. He tried to counter the arguments against a moving earth as best he could, but his arguments were not convincing because a heliocentric system cannot be understood properly without getting rid of Aristotelian physics. This point was well understood by Copernicus' successors if not by Copernicus himself.

In addition to the physical question of whether the earth moves, there was the metaphysical question of what it means for the earth to move (the cosmological level of Copernicanism). The medieval universe was small and cozy. Everything is in its proper place and nothing important ever changes. You couldn't tear the earth out of its natural place in the

apparent path of planet
as seen from earth

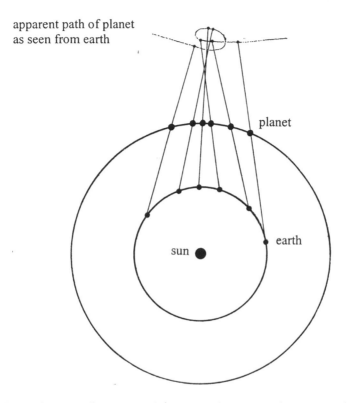

planet

sun

earth

Figure 3. A schematic illustration of the reason for retrograde motion. The path of the planet, as seen from the moving earth, appears to go backward during part of the orbit.

center of the cosmos, send it whirling around the sun, and still have everything be basically the same. Copernicus had given metaphysical reasons why the sun should be at the center of the cosmos, but that did not solve the problem. Once change and instability were allowed by the introduction of heliocentrism as a viable alternative to the traditional universe, there was no way to keep further novelties from being invented and proposed. As this became more and more apparent to the authorities of the Church, they hardened their initially tolerant attitude toward Copernican astronomy. But by then, it was too late to go back to the old thinking.

The importance of Copernicus was not the result of his particular mathematical description, which was incorrect. Nor was it the result of his explanation of his system, which was based on old and incorrect ideas. The importance of Copernicus arose from the fact that the publication of his system started an important debate and led to further advances in thinking. Copernican astronomy served as a springboard for progress both in mathematical astronomy (by Kepler) and in the science of motion (by Galileo). In a final irony, Copernicus almost surely would have detested the new ideas in both of these areas.

§3. KEPLER AND TYCHO

Johannes Kepler was a mathematician of genius, and he was thoroughly steeped in Pythagorean number mysticism. His Pythagorean outlook induced him to look for harmonies and numerical relationships in the orbits of the planets; his genius, combined with patience, allowed him to find these relationships. But Kepler also had one more trait that made him different from other intellectuals of his time: he insisted that the relationships he found must describe the actual paths of the planets. Those which did not were unacceptable. This may sound trivial to those of us in the modern world, but it was a revolutionary attitude in his day. This attitude in combination with his mathematical talent and his mystical predilection culminated in the crucial discoveries that astronomy needed at that time.

In his youth, Kepler thought he had discovered a relationship between the orbits of the five planets (in Copernicus' system) and the five Platonic solids (see chapter 18). He later abandoned this idea because the numerical values of the orbits were not quite right. He remained convinced, however, that deep numerical and mathematical relationships existed in the cosmos, if only he could find them. But as Kepler continued his work, he became increasingly aware that he was hampered by the low quality of existing astronomical observations. By a stroke of good luck, new observations were being made at that time, far better than any which had ever been made before. These new observations were the work of the greatest observational astronomer since antiquity, the Danish nobleman Tycho Brahe. Brahe had a passion for systematic and precise measurements, and he had the resources to build new instruments of his own design in order to make such measurements. Not only were his planetary positions recorded with the highest precision attainable (until the telescope came into use), they were also recorded for many points in the orbit instead of just a few. Tycho's greatest accomplishment was the set of planetary observations that Kepler later used so brilliantly. In addition, Tycho observed the paths of comets and showed that comets were definitely celestial objects that went through the orbits of the planets, proving that the crystal spheres of Aristotle must be fictional. Kepler joined Brahe at the Imperial court in Prague, thus gaining access to the planetary measurements (which had never been published). When Tycho Brahe died, Kepler was able to keep the precious measurements in his possession.

Kepler worked on the problem of planetary orbits for the next several decades, checking his ideas and calculations against Tycho's recorded positions. The planet that gave him the most trouble was Mars, and for a good reason. Like Copernicus and Galileo, like Plato and Aristotle, like all astronomers before him, Kepler assumed that the planets moved in circles. But as we now know, the planets don't move in circles. They move in ellipses (an ellipse is a kind of flattened circle; the orbits of planets are only slightly elliptical, almost circular). Different planets depart from circularity by different amounts. Mars, it so happens, has a greater departure from circularity than most planets. So when Kepler tried to find a circle that fit the elliptical orbit of Mars, he could not do it. He tried and tried, doing laborious calculations by hand for years, using one circular scheme after another. He came very close, but because Tycho's observations were so precise, close was not good enough. In desperation, Kepler gave up the assumption of circular paths and discovered that he could get a perfect fit with an ellipse. This extremely important discovery, that planets move in elliptical orbits, we now know as Kepler's first law.

During the same investigation of Mars' orbit, Kepler also made another important discovery. Everybody had always assumed that the planets move with a constant speed. It became apparent to Kepler that the speeds of the planets actually change, moving faster when they are closer to the sun and slower when they are farther away. This was very disturbing; the planets should move in a more orderly and harmonious fashion than that. Kepler found through his difficult and tedious calculations that there was indeed an underlying mathematical harmony hidden deep in the orbits. Kepler's discovery is illustrated in Figure 4 (the departure from circularity is exaggerated compared to real orbits for purposes of illustration). The pie-shaped wedges are marked out by movements of the planets. In both cases, the planets took equal amounts of time to move the distances shown. The wedge for the planet near the sun is a little wider because the planet is moving faster. What Kepler discovered is that the sizes of the two wedges, that is, the crossed-hatched areas, are both equal. Any wedge marked out by the same time interval is also equal in area. We now call this Kepler's second law: equal areas are swept out in equal times. I don't know whether you find this interesting or amazing, but to Kepler's Pythagorean mind it was like finding the signature of God among his numbers.

A final mathematical relationship discovered by Kepler concerns the time it takes for a planet to go around the sun and the planet's distance from the sun. The planets move slower and slower as they get farther and farther from the sun. Is there some order or reason to this slowing-down? Yes! Kepler discovered that the square of the time a planet takes to go

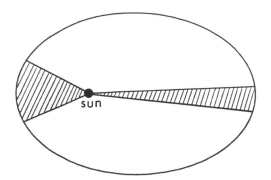

Figure 4. An illustration of Kepler's second law. The crosshatched wedges represent areas swept out by moving planets (this ellipse is more exaggerated than a real planetary orbit, but is somewhat like a comet's orbit). The areas of the two wedges are equal, and they are both swept out in equal amounts of time.

around the sun is proportional to the cube of the planet's distance from the sun. This is a complicated relationship; we can only imagine how Kepler extracted it from his reams of numbers, and we can only imagine his joy at finding once again a hidden harmony in God's plan for the cosmos. We now call this relationship Kepler's third law.

Kepler's work is important for three different reasons. The first reason is that Kepler finally provided a correct mathematical description for the motions of the planets (this is the point usually emphasized in modern textbooks). After thousands of years of effort, the true paths of the planets were finally known. Moreover, this correct description was needed before Newton could make further progress by finding the explanation for these orbits (we'll discuss Newton later). A second reason for the importance of Kepler's work is that he chose to give up his beloved circles because they were contradicted by Tycho's observations. No one before Kepler would have given up their metaphysical presuppositions just because the data didn't fit. In this sense, Kepler represents a transition to the modern view of science. His Pythagorean number mysticism was a relic of the past, while his insistence on agreement with the evidence of the senses was a jump into the future. Finally, Kepler's third important contribution is that he thought about the motions of the planets in terms of physical causes. As long as everyone assumed the planets move in perfect circles and that perfect circular motion was a natural motion, then there was no need to ask any more questions. But when Kepler discovered that the planets move in elliptical paths (having the properties given by his laws), he felt that these discoveries did require an explanation. He knew that the sun must have a central role in directing the motions of the planets, and he imagined some sort of physical cause emanating from the sun that weakens as distance from the sun increases. As the cause becomes weaker, the planets at an increased distance move more slowly. This combination of physical cause with mathematical description has proven to be a remarkably fertile idea in the development of science. Because Kepler was exploring completely new intellectual territory, he had many of the details wrong. But his key insight, namely that something from the sun causes the planets to move as they do, was profoundly correct.

§4. GALILEO AND DESCARTES

We now take a look at the role played by Kepler's brilliant contemporary, Galileo Galilei. Galileo probably did more than any other thinker to bring about the establishment of modern science. Like Kepler, Galileo believed that mathematics is the language we need in order to understand nature. He was undoubtedly influenced in this belief by the same Pythagorean tradition of Neo-Platonic thought that influenced Kepler. But Galileo had a more practical and modern mind; he was much less prone to mystical speculation. (Another irony: the hardheaded Galileo continued to believe in circles while the otherworldly Kepler gave up circles to match the observations.) Galileo also championed experimentation as a way to learn about nature, and it was his combination of experimentation and mathematics that proved to be so powerful in advancing physics.

Galileo did not make any major contributions to mathematical astronomy, but he made three other important contributions to understanding the planets: telescopic observations, a new science of motion, and popularizing Copernicanism. The most important contribution turned out in the end to be the new science of motion he invented. During his lifetime, however, he was most famous for his work in astronomy, using the recently invented telescope. Galileo significantly improved the telescope, and he was the first person to point it toward the night sky and discover new things there. He discovered the mountains and craters of the moon, proving wrong the Aristotelian claim that celestial orbs were perfect spheres. He discovered that Venus had phases like the moon, important evidence that Copernicus was right. He discovered that another planet, Jupiter, had moons of its own. This exciting discovery provided a miniature Copernican system for people to see. Galileo's third contribution was to convince the majority of educated people, who still mostly lived in a geocentric universe, that Copernicus was right.

Making a convincing case for Copernicanism is probably what Galileo is most known for today. His rhetorical skills were formidable, and he wrote in strong and interesting Italian prose instead of boring academic

Latin. The popularization of heliocentrism was important in advancing the cause of science at the time, especially since many of Galileo's arguments were based on the inadequacy of Aristotle's physics. Galileo's new thinking about motion made the often-quoted arguments against a moving earth obsolete. The old physics, along with the old cosmology, needed to be swept away in order to make progress. Galileo's insistence that the same laws of nature hold on earth and in the celestial realms was particularly important. Change does not come easily, however, and Galileo is also famous today for the reaction against him. A brief digression on this battle is worthwhile. The initially tolerant (and even encouraging) attitude of the Church toward Copernicanism had gradually hardened into opposition, as it became apparent that new ideas in astronomy might lead to new and heretical religious ideas. The entire intellectual climate of Europe had become less tolerant due to the religious wars and persecution resulting from the struggle between Protestants and Catholics. Also, Galileo's intellectual enemies, the Aristotelian academics, had worked hard to convince the Church to side with them. Galileo, for his part, entered into needless disputes and controversies with influential members of the clergy (although brilliant, he was also abrasive and egotistical). All of this culminated in the well-known trial of Galileo, in which he was forced to deny Copernicanism, and the Church rejected heliocentrism for centuries.

In the last few years of his life, Galileo wrote his masterpiece, a systematic explanation of the new science of motion that he had developed over his lifetime. Almost every student of high school or college physics learns this work at the beginning of the course: the relationships between position, time, velocity, and acceleration. Galileo's new science was one of the crucial ingredients, along with Kepler's mathematical astronomy and ideas about physical forces from the sun, that Newton synthesized into a genuine understanding of planetary motion. The remaining ingredient that Newton required was supplied by the French philosopher, René Descartes. Descartes was an immensely influential thinker in both philosophy and mathematics (most of his work is outside the scope of our interest here). Descartes' physics largely turned out to be wrong, except for the idea of inertia. Descartes' concept of inertia was both correct and important. The old Aristotelian concept of inertia was that the natural state of motion is rest, and objects only move if something moves them. Galileo correctly denied this, but incorrectly thought that the natural state of motion was circular. Descartes finally realized that the natural state of motion is to go in a straight line with a constant speed. (This conclusion is by no means obvious because we almost never see it happen; there are always forces acting that prevent the natural state of motion from occurring.) So, inertia is the property of an object that keeps it going in this natural straight-line motion. Beginning physics students learn this assertion as Newton's first law of motion.

But if the natural thing to do is to fly off in a straight line, why then do the planets move in ellipses around the sun? This is the part Descartes had wrong. Unable to accept Kepler's ghostly emanations from the sun, Descartes thought that some substance must fill the universe and account for the motions of the planets. He imagined that this substance swirled around in cosmic whirlpools, carrying the planets along with it. Kepler was closer to the truth in some ways, thinking there must be a force from the sun that weakens with distance. But Kepler had the wrong concept of inertia; thinking that the planets wanted to be at rest, he imagined the force from the sun must push them along to keep them moving. Of course, Galileo had it wrong, too. His concept of inertia was circular motion, and he thought that the planets moved in circles, making further explanation unneeded. It was Newton who put all the pieces of the puzzle together in just the right way to explain the motions of the planets properly.

This collection of half-right and half-wrong theories may be hard to follow, but following each twist and turn of the argument isn't really necessary. I included all this detail because I have a point to make, and my point is this: Scientists don't simply sit down and write out correct theories from scratch. Ideas evolve over time. Different people contribute different bits and pieces, good ideas are mixed with bad ideas, observations are improved, new ideas are invented, and the good ideas are slowly sifted out from the ideas that don't work. Different combinations of good ideas are tried, and eventually a coherent picture emerges. In this case, it emerged from the mind of Newton.

§5. NEWTON

Isaac Newton's accomplishments are staggering. He did fundamental work in optics that dominated the field for hundreds of years; he invented the branch of mathematics known as calculus; and he founded physics as we know it today by creating mechanics, the science of motion. In the course of creating mechanics, Newton discovered the correct explanation for the motions of the planets. Newton and his colleagues had by this time refined Kepler's nebulous concept of "a force from the sun" into "the force of universal gravitation." The force of gravitation was a mutual attraction between all things that obeyed a well-defined mathematical form (namely, that decreased as the inverse distance squared). Building on Galileo's study of acceleration, Newton discovered the general way in which a force changes the motion of an object (we now call this discovery Newton's second law of motion). Applying his new mechanics to planets attracted by the gravity of the sun, Newton was able solve the ancient problem of planetary motion.

In words, the solution is this: A planet would just travel in a straight line if left on its own. But the planet is not left on its own, it is pulled toward the sun by gravity. On the other hand, if the planet were just standing still to start with, gravity would cause it to fall straight into the sun. So, a planet has two opposing tendencies. The planet has an inertial tendency to fly off in a straight line, and it has a gravitational tendency to fall into the sun. The delicate balance of these two opposing tendencies results in the planet's elliptical orbit. Newton showed mathematically that his laws of motion, applied to an inverse-distance-squared attractive force like gravity, predicted all of the properties that Kepler had discovered in the planetary motions (i.e., Kepler's laws). The beautiful mathematical harmonies of Kepler turned out to be the result of an even deeper mathematical order. Newton's laws of motion, which we now call classical mechanics, could explain and predict any motion due to any kind of force. This theory was applied with unchallenged success for centuries (see chapter 4) until it too was found to have limits. But that is a whole new story. The story of the motions of the planets, which started in ancient Babylon, ended with Newton.

§6. COMMENTS

What are some of the interesting points concerning scientific discovery that are illustrated by this story? One point is that merely making observations and measurements is not enough for a science to evolve; people must also ask questions and look for explanations. This is why astronomy as a science was founded by the Greeks and not by earlier civilizations. The other side of this coin, however, is that science does need better observations and more precise measurements in order to make progress, particularly if conditions are ripe for a major breakthrough but the necessary data are lacking. We saw an example of this in Kepler's utilization of Tycho's work.

A theme running through the entire story is the influence of metaphysical ideas on the thinking of scientists. The assumption of circular motion was common to Ptolemy, Copernicus, and Galileo. The joining of Pythagorean heliocentrism with dogmatic adherence to uniform circular motion was the foundation of Copernicus' system. And it was the combination of a Pythagorean belief in mathematical harmony with a respect for the precise measurements of Tycho that inspired Kepler's work. Another theme in this story is the influence of social and cultural forces on the history of ideas. The humanist activity and the ferment of new ideas in the Renaissance exposed Copernicus to heliocentrism at a time when the Ptolemaic system was starting to be perceived as weak. The practical need for better astronomy (navigation, calendar reform, and so forth) coincided with a general reevaluation of old ideas. This reevaluation included the realization that Ptolemy's work gave inaccurate predictions, which could only be patched up by adding more epicycles. Geocentrism was becoming a tired and decaying system. So, when Copernicus proposed his heliocentric system, it started a revolution in thought instead of being quietly forgotten as Aristarchus' system had been forgotten.

FOR FURTHER READING

The Origins of Modern Science 1300–1800, by H. Butterfield, Macmillan, 1960 (originally G. Bell & Sons, 1949).
The Scientific Revolution 1500–1800, by A. Rupert Hall, Beacon Press, 1954.
The Sleepwalkers, by Arthur Koestler, Grosset & Dunlop, 1959.
The Scientific Renaissance 1450–1630, by Marie Boas, Harper & Row, 1962.
Thematic Origins of Scientific Thought, by Gerald Holton, Harvard University Press, 1973.

EPIGRAPH REFERENCE: Gerald Holton, *Thematic Origins of Scientific Thought*, Harvard University Press, 1973, p. 70.

PART II

MENTAL TACTICS: SOME DISTINCTIVELY

SCIENTIFIC APPROACHES TO THE WORLD

Chapter 6

A UNIVERSE IN A BOTTLE: MODELS, MODELING,
AND SUCCESSIVE APPROXIMATION

> One of Oppenheimer's great strengths as a theorist was an
> unerring ability to look at a complicated problem and strip
> away the complications until he found the central issue
> that controlled it.
> *(Kip Thorne)*

IN TRYING to understand nature, we rarely attempt to grasp completely every possible detail. If we did, we'd be overwhelmed by the mass of inconsequential information. As a result, we would miss the truly interesting patterns and relationships that give us scientific insight. An important tool to achieve scientific understanding is the construction of conceptual models. Models, in the sense in which I'm using the word here, are imaginary simulations of the real natural systems we are trying to understand. The models include only properties and relationships that we need in order to understand those aspects of the real system we are presently interested in. The rest of the details of the system are left out of the model.

§1. BILLIARD BALLS AND STREET MAPS

Let's illustrate this vague and abstract description of models with a brief example. Many properties of gases can be understood by modeling the gas atoms as tiny billiard balls; in other words, the atoms can be imagined as small hard spheres that don't affect anything unless they collide with it. When they do collide, according to the model, they behave similarly to colliding billiard balls (the similarity is in the way they exchange physical properties like momentum and energy). Now, atoms are not tiny billiard balls. Atoms have many properties that are not at all described by this model (think of chemical reactions, for example). And yet, many properties of gases (e.g., pressure, heat capacity, expansion and compression, diffusion, etc.) can be understood on the basis of this very simple model.

A good example of a model from everyday life is a street map. The lines on paper are not real streets, and yet the map models many properties of the real streets: the relative positions of the streets; where the streets intersect with each other; the directions of the streets; and so on. Most of these properties of streets are in fact the properties we are interested in when we use a map. Many other properties are not provided by our map model: the people walking on the sidewalks; whether the street is asphalt or concrete; whether it runs through a business district or residential area; and so on. Many of these properties are of no interest when we use a map, though it's possible that some piece of information that you want might be left out. A map that includes most of the information you want, and leaves out most of the information you don't, is a good model of the real streets for your purposes. Whether a model is good or not depends on what you want the model to do. The worth of a model depends on what kind of information or understanding you wish the model to provide. In the case of the map, a good map for navigating the downtown region of a city will be different from a good map for getting to another city hundreds of miles away. This point is also true for scientific models. The billiard ball model for gas atoms is a good model for explaining gas pressure, but it's a poor model for explaining how a gas condenses into a liquid.

Sometimes, a model just isn't appropriate for understanding a process or system we are interested in. For those cases, we need to devise a new model. But sometimes, we don't need to invent a whole new model. We can modify, extend, or refine a model so that it allows us to understand things the unmodified model did not. For example, maps sometimes model freeway exits in a very sketchy and schematic way, showing no details of how the exit is actually laid out in real life. This might be alright in some cases, but I have occasionally had a lot of trouble getting on or off a freeway due to this mapmaking custom. When this happens, I definitely regard the model as deficient. Yet the model (map) can easily be fixed in this case simply by adding the details of the freeway exit to the existing map. In other words, the sketchy map is an approximate model of reality that is too crude; the modified map is an improved approximation. This idea of approximate models and of successively improving your approximations (until your model does what you want) is very useful in the sciences. For example, we have identified a shortcoming of our billiard ball model of gases (it doesn't explain why a gas should condense into a liquid). Now that we know where it fails, we can refine the model to take care of this problem (e.g., we can endow the billiard balls with a new property like mutual attraction).

We'll come back to this gas example in more detail later, as well as a number of other examples from the sciences. Before we look at more examples, however, we need to consider one more general aspect of scientific models: the value of mathematical models. We have so far discussed the idea of a model as a pictorial representation of reality, and sometimes this is an appropriate way to make progress and gain new understanding. Sometimes, however, a conceptual model is only a first step, and the second step is a mathematical representation of the conceptual model. In the scientific examples of §2, I will try as much as possible to indicate how this is done and why it's useful (though a detailed mathematical treatment of those examples is beyond the scope of this book).

Traffic Model

Let's end this section with an example of a fairly simple mathematical model. The system we'll model is a familiar one: traffic flow at an intersection governed by a traffic light. Suppose we are interested in the traffic at an intersection of two streets, Main Street and Oak Street. Main Street is usually a little busier than Oak Street, but during rush hours Main Street becomes much busier. Now, there's a traffic light at the intersection, and we want to arrange the timing of the red and green lights so that the number of people inconvenienced by stopping is about equal on the two streets. Clearly, in order to accomplish this goal, the people on the busier street must get a shorter red light. The difference between rush hours and the rest of the day makes solving our problem more complicated. We might ideally want a computer-controlled traffic light that adjusts its timing to the traffic flow, but the intersection is located in Budgetville where the residents are too cheap to buy one. So how should we arrange the timing to accomplish our goal? We need to equalize somehow the number of people inconvenienced on the two streets over the course of an entire day. The solution to our problem will involve some kind of compromise in which people on Main Street will be somewhat more inconvenienced during rush hours while people on Oak Street will be somewhat more inconvenienced the rest of the time.

In order to progress further, we'll make an idealized mathematical model of the situation. We said that Main Street is busier than Oak Street, more so during rush hours. To make a mathematical model, we need a mathematical concept that corresponds to (and measures) how busy a street is. The rate of traffic flow, that is, the number of cars passing by per unit time, is a good candidate for this measure. All of the complications of changing traffic flow rates throughout the day can be approximated by four numbers, two for Main Street (rush hours and nonrush hours) and two for Oak Street (rush hours and nonrush hours). Choose symbols to stand for these four different rates.

$$R_{m,r}, R_{m,n}, R_{o,r}, \text{ and } R_{o,n}$$

R stands for rate of traffic flow, the m subscript stands for Main Street, the o subscript stands for Oak Street, the r subscript stands for rush hours, and the n subscript stands for nonrush hours. $R_{m,r}$, for example, is the traffic flow rate on Main Street during rush hours. Numbers like these in a model are sometimes called the parameters of the model.

We are now ready to solve our model. Remember, we have no control over the rates; they are determined by the habits of the Budgetville citizens. We can measure them, but not change them. The variables that we can control are the times that the lights are red for each street. What do we want? We want equal numbers of people inconvenienced by stopping on each street. We want to arrange the timing of the red lights so as to make this equality true. Finding these correct times for the red lights is what we mean by solving the model. Let's call the duration of the red light times Δt_m on Main Street and Δt_o on Oak Street. A little thought will convince you that we now have four different values for the number of cars stacked up at red lights. These are

$$N_{m,r} = R_{m,r}\, \Delta t_m$$

$$N_{o,r} = R_{o,r}\, \Delta t_o$$

$$N_{m,n} = R_{m,n}\, \Delta t_m$$

$$N_{o,n} = R_{o,n}\, \Delta t_o$$

where $N_{m,r}$ is the number of cars lined up on Main Street during rush hours, and so on for the other three numbers.

The crucial question now is this: How are these numbers related to our goal? We want equal numbers of people on the two streets inconvenienced by having to stop. The total number of people stopped on Main Street or Oak Street during rush hours is equal to the number stopped during each red light multiplied by the number of times the light turns red. The same can be said about the total number of people stopped during the nonrush hours. Assume that rush hours are 7:00 A.M. to 9:00 A.M. and 4:00 P.M. to 6:00 P.M., a total of four (4) hours. Nonrush hours are 9:00 A.M. to 4:00 P.M. and 6:00 P.M. to 9:00 P.M. (we'll assume there's negligible traffic after 9:00 P.M.; there's not much nightlife in Budgetville), a total of ten (10) hours. Since

$$10/4 = 2.5$$

we can write our equality condition in the form

$$N_{m,r} + 2.5N_{m,n} = N_{o,r} + 2.5N_{o,n}$$

This is a mathematical statement of our desire that equal numbers of people be inconvenienced throughout the day on both Main and Oak Streets. Now we are getting somewhere! Rewrite this condition as

$$R_{m,r}\, \Delta t_m + 2.5R_{m,n}\, \Delta t_m = R_{o,r}\, \Delta t_o + 2.5R_{o,n}\, \Delta t_o$$

and find Δt_m in terms of Δt_o. If we do the algebra, we discover that

$$\Delta t_m = \{(R_{o,r} + 2.5R_{o,n})/(R_{m,r} + 2.5R_{m,n})\}\, \Delta t_o$$

We have now accomplished our goal. Knowing values for the four R parameters, we can now set the times of the lights using this expression. Doing so will at least approximately yield the desired result: equal numbers of people stacked up during the day at red lights on Main and on Oak.

This example illustrates several interesting points about models. A number of idealizations and approximations were made in order to create a mathematical model of the real-life situation, and using the model enabled us to extract certain kinds of information that would have been difficult to obtain otherwise. The model employs a number of parameters, which approximately represent some quantities of interest in the model. From our model, we don't get any new insight or understanding into certain questions and issues (e.g., we don't learn anything about why Main is busier than Oak); these questions may be of interest in some other context, but our model is not intended to address them. Finally, we can test our model empirically to find out if the approximations we've made are close enough to reality to suit our purposes: Any complaints in Budgetville?

§2. Some Scientific Models

To see how models are·used in the sciences, let's look at some examples of models that provide insight into various natural systems and processes. These brief discussions are only intended to give a sense of how the models work, without going into many details. I don't give any mathematical formulations of the models, but I do indicate the ingredients needed for such formulations. The examples we will look at are the ideal gas model; models of blood flow in the body; the nuclear shell model; models of drug uptake in the body; biological models of heredity; and the game theory model of social conflict.

Ideal Gas

Let's take another look at the model of gas molecules as billiard balls, which is technically known as the ideal gas model. In this model, the gas molecules are pointlike, that is, they don't take up any space. The molecules don't interact with each other or with anything else (e.g., the walls of their container) unless they happen to collide (make a direct hit). If they merely pass nearby, it's as if they weren't even there. When they do collide, they exchange energy and momentum as particles undergoing a classical elastic collision. Which is to say: they act like colliding billiard balls, or like tennis balls bouncing off a brick wall. These properties define the molecules of an ideal gas. Ideal gas molecules are a good approximation to the behavior of real gas molecules when the molecules are far apart and moving fast, so the ideal gas model works best for gases with a low density and high temperature. The ideal gas approximation breaks down when the density is too high or the temperature too low, because other ways in which the gas molecules interact (e.g., they attract each other) then become more important.

This model can be used to explain the relationships between pressure, volume, temperature, and number of molecules. In other words, we can explain why the ideal gas equation (see chapter 19) is true, using our model. The key idea is that pressure is caused by gas molecules hitting the walls of the container. (This idea can be treated mathematically by relating the pressure on the wall to the momentum imparted by the molecules hitting the wall.) If the volume of the container is increased, the gas atoms become more spread out, and the pressure decreases. If the temperature is increased, the gas atoms move faster, and the collisions with the walls become harder and more frequent, thus increasing the pressure. The qualitative relationships found in the ideal gas equation are obviously predicted by our model, and it turns out that the quantitative relationships are also predicted by the mathematical version of the model.

Common sense tells us why the model must break down at too high a density. According to the ideal gas equation, the product of pressure and volume is constant. If this remained strictly true, we could make the volume as close to zero as we pleased by simply continuing to increase the pressure. This conclusion makes sense in terms of our model because the gas molecules in the model don't take up any space. But the model itself doesn't make any sense if the molecules get too close together, because in fact they do take up space. The approximation (pointlike molecules) breaks down when the density becomes too high, and so does the ideal gas equation. If we did the experiment, we would at first start to see small deviations from the equation as the pressure increased. Eventually the gas turns into a liquid when the pressure becomes high enough. A liquid obviously doesn't behave like an ideal gas! More particularly, the volume of a liquid decreases very little as the pressure increases.

In the range of densities and temperatures for which the deviation from ideal gas behavior is still small, we can retain our model as a first approximation. We just modify the model slightly to account for the deviations. For example, we can add a weak attractive force (which decreases with distance) to account for the tendency of the gas to form a liquid, and we can add a strong repulsive force at very short distances to account for the nonzero size of the molecule. Various mathematical versions of such improved models have been proposed and work well, accounting for many of the behaviors we see experimentally. Eventually, however, we discover behaviors (the gas mixture of oxygen and hydrogen turning into water, for example) that cannot be explained by any amount of tinkering with the model. To account for these cases, we need a completely different conceptual model of the gas molecules.

Blood Flow

Understanding how our blood flows in veins, arteries, and capillaries is an important problem of great medical interest. But this problem is also incredibly complicated to understand completely. As usual, we'll begin our attempt to understand a complicated situation by using a simple model. We'll start by approximating blood as a liquid with the properties of water. Next, approximate the blood vessel as a thin straight tube with rigid walls and a uniform cross-sectional area. Finally, assume a moderate steady pressure on the liquid. This model is a good approximation for water flowing in a hypodermic syringe, for example. It's admittedly a poor model for blood in the human body, but has the virtue of being easy to solve. The velocity of the blood flow in this model is directly proportional to the pressure drop along the length of the blood vessel.

The first incorrect assumption made in this model is that the resistance to blood flow (which is the constant of proportionality in the model) is a constant. The resistance of the blood vessel to blood flow depends on the radius of the vessel, and the radii of the small arteries can be greatly varied by their muscles. But the pressure drop itself can also vary as the radius of the vessel changes. Our initial model may be a good starting point, but a much more complicated model is required to accurately describe blood flow throughout the circulatory system. The second problem with our model is the assumption of flow through a straight tube. Blood vessels are generally curved tubes. Worse yet, the arteries are continually branching into numerous smaller tubes, down to the capillaries. In these geometries, fluid flow may become very complicated due to the onset of turbulence. The mathematical models needed in such cases are very difficult to solve, requiring large amounts of time on high-powered computers. A third problem with our model is the treatment of blood as a simple fluid, like water. If all we had was blood plasma, then this assumption might be reasonable. The chemical complexity of blood plasma doesn't affect its fluid properties much. But the blood cells suspended in the plasma make its fluid properties much different (and more complicated) than water. Because of these cells, the viscosity (resistance to flow) of blood actually depends on the flow rate of the blood. The reasons for this effect are not well understood, but it's certainly important; the smallest capillaries are barely bigger than the red blood cells. To model the actual biological effects in this case is probably not practical. If we assume some fairly simple model for the variation of the viscosity, however, we can still approximately find the blood flow.

We may also need to consider some other features of blood circulation (depending on the particular problem we are considering). The pressure, for example, isn't steady but rather comes in big pulses due to the heartbeat. These time variations in pressure become less prominent in the smaller vessels, which are farther away from the heart. Blood vessels expand to accommodate blood flowing into them, which keeps the pressure from getting too high (age and cholesterol decrease this ability). Another complication is that the circulatory system is equipped with a set of valves to shunt blood into or out of the capillaries as needed.

Comparing this example (blood flow) and the previous example (ideal gas), we see an interesting difference. The simplest model for the gas works well and only needs small corrections under certain conditions. The corrections only become large as the conditions become extreme. In contrast, the simplest model for blood flow isn't really very good under any conditions. The utility of that simple model is that it provides a framework to think about the problem and to ask which improvements are needed. The answer to the question of what improvements are needed depends in turn upon which features of blood flow we are interested in understanding (and under what conditions). The improvements in this case will not be small corrections but instead will radically change the simple model.

Nuclear Shell Model

The nucleus of an atom is made up of protons and neutrons. During the 1930s and 1940s, after the discovery of these subatomic particles, physicists worked to understand the properties of nuclei in terms of the forces holding the particles of the nuclei together. One model assumed that the particles lost their individual identities because each particle was acted on by strong forces exerted by all of the other particles in the nucleus. In this model, called the liquid drop model, the nucleus is thought of as a kind of homogeneous substance rather than as a collection of single particles. A different model, called the shell model, assumed that the particles do retain their individual identities. As single particles feeling some kind of force, the particles in a nucleus might behave similarly to the electrons in an atom. Now, the electrons in an atom are in "energy levels," also known as orbitals or shells. The organization of these electron shells accounts for the structure of the periodic table of the elements (see chapter 2) in chemistry. But the electron shells exist because there is a single point source (the nucleus itself) for the forces on the electrons. No such single source had been determined for the forces on the particles in the nucleus, and for this reason the nuclear physicists heavily favored the liquid drop model.

In her examination of the data on stability and abundances of nuclear isotopes, a physicist named Maria Goeppert Mayer discovered an interesting pattern. (Isotopes are forms of an element that have different masses, i.e., different numbers of neutrons.) The pattern was that nuclei with certain special numbers of neutrons or protons (namely, 2, 8, 20, 28, 50, 82, and 126) were unusually stable. Now, atoms with certain special numbers of electrons are also exceptionally stable. These atoms are the inert gases, and they are stable by virtue of their "closed shells" (a closed shell means that the energy level is filled with all the electrons it can accommodate). Mayer reasoned that the so-called magic numbers for the nuclei might also be explained by such a shell model (J. H. D. Jensen also independently devised the shell model at about the same time). The forces between particles in a nucleus were unknown at that time, so Mayer started by using a simple model in which the particles move freely inside the nucleus but can't escape from it (this model, well known to physics students, is called a square well). This model predicts the first three magic numbers (2, 8, 20) but then breaks down and cannot account for the numbers above 20. We seem to be on the right track, but there is clearly something important going on in the real nucleus that is missing from the model. Mayer tried a number of modifications to the model, initially without success.

One aspect of the problem she had been leaving out of the model was a term analogous to the magnetic energy of the nuclear particles. The protons and neutrons can be thought of as similar to little compass needles, and they have different energies depending on their orientations just as a compass needle lowers its energy by pointing north. This effect, known technically as a spin-orbit interaction, occurs for electrons in atoms and was well understood for that case. But because the effect is very small for electrons in atoms, no one had yet thought of including it in the model of the nucleus. These energies turn out to be large in the case of the nucleus, however, and significantly modify the results of the initial simple model. Mayer calculated the predictions of a square well model including spin-orbit terms, and this improved model correctly predicted all of the magic numbers! Shell models became an accepted part of nuclear physics. Since that time (1949), a great deal has been learned about nuclear forces, and we now have considerably more refined models. But the basic framework of these more refined models is the idea of shell structure, which was demonstrated using the improved (but still simple) model just discussed. In that sense, the use of a relatively crude model made a decisive and fertile contribution to our knowledge. Once again, we needed to have the essential features (both the square well and the spin-orbit terms) but not all of the details. Note that we started with a model that was inadequate (though basically correct) and looked for the necessary ingredient missing from this model in order to explain the facts of interest to us at the time (the magic numbers). Incidentally, Mayer (who was not paid a salary for much of her career due mostly to gender discrimination) later shared the Nobel Prize for this work.

Drug Uptake in the Body

Another interesting biomedical problem is the way in which drugs, typically eaten as a pill or injected into the bloodstream, make their way to the part of the body affected by the drug. An anticancer drug might have to get to the tumor it's intended to combat, for example. The actual physiology involved in this process is very complicated. The drug will have varying concentrations (which all change with time) in the bloodstream, in the fatty tissue of the body, in various organs (e.g., the liver), and in the part of the body we have targeted. The mechanisms by which the exchanges between different parts of the body take place, and how these mechanisms depend on the concentrations, are not well understood. But if we are interested primarily in more practical questions (such as the time it takes for the drug to reach its target, the amount that reaches there, and the best method to administer the drug), then we may not need to understand all of these mechanisms. We can instead make a simplified model. The drug concentrations can be assumed, as an approximation, to change with time in some fairly simple fashion (e.g., exponential; see chapter 20). The parts of the body (blood, fat, organs, etc.) can be modelled as just a number of storage areas for the drug, and the rate at which the drug is exchanged between these areas is then all we need to know. A highly oversimplified model like this one is not intended to provide insight into the workings of the system it models. The intention in this case is merely to answer some limited questions about the behavior of the system. Now, since we know little about the physiological processes involved, we can't predict the exchange rates used in our model. A practical procedure in a case like this is to make measurements of some quantity that is accessible (in this model, the drug concentration in the blood is a good example of an easily measured quantity). This measured data is then combined with the model in order to figure out the numerical values of the exchange rates. Using these rates in our model, we can then predict (at least approximately) the drug concentration in other parts of the body at future times.

We have here another example of using models to answer particular questions rather than to investigate the fundamental workings of the system we are modeling. In this sense, the drug uptake model is more similar to the traffic model than to the other scientific models we've examined. Models are used this way extensively in various applied sciences. Can we use such a model to learn anything interesting about nature? Yes, sometimes even highly oversimplified models can provide us with real scientific insight. For example, if one model works and another model doesn't, we have a clue about what to include in a general theory of the system. Also, parameters in a model (such as the exchange rates in our drug uptake model) may correspond to real quantities that we use to conceptualize the mechanisms of the system (in this case, physiological processes). We can then use the numerical values of these parameters to test predictions of more sophisticated attempts to understand the system.

Notice the interesting difference between parameters in the drug uptake model (exchange rates) and parameters in the traffic model (traffic flow rates). In the traffic model, we can measure the parameters directly (just sit at the intersection with a stopwatch and count the cars). The numerical values of the parameters are then simply put into the model. In the drug uptake model, however, we can't just measure the exchange rates; they aren't accessible to our measurement techniques. Instead, we measure the change with time of one variable (drug concentration in blood), and then look for values of the rates that force the model predictions to agree with these measurements. This is called fitting the parameters of a model, a procedure used extensively in the sciences. We'll come back to the topic of parameter fitting in §3.

Heredity

Genetics is the science concerned with the transmission of hereditary information. We can think of genetics as a series of successively more sophisticated models. We've now looked at several cases in which successively more refined models (we might call them closer approximations) replaced simpler models. In the case of genetics, each of the models we'll discuss was devised to (and was able to) account for all of the information known at the time it was developed. Each improvement was inspired by new experimental information. So in this case, the series of models is also chronological. The earliest model is that of Gregor Mendel, the well-known founder of genetics. In this model, there are dominant traits and recessive traits that are passed on to later generations according to the rules that Mendel deduced from his pea plant experiments. If the traits (tall/short, wrinkled seed/smooth seed, etc.) are dominant, they are expressed when received from either or both parents; if the traits are recessive, they are expressed only when received from both parents. The key concept in this model is the trait itself, and there is no attempt to conceptualize the physical mechanism by which the trait is passed on. There is only some factor of unknown nature postulated that must somehow carry this information. By the time Mendel's work became generally known to scientists, cell biologists had observed, through their microscopes, threadlike structures called chromosomes in the nuclei of cells. The chromosomes were identified as the carriers of hereditary information, and this discovery was combined with Mendel's work to create a new model. An important implication of this new model, verified by experiments, is that some traits are linked with other traits, that is, they are inherited together. This linkage is due to the traits being carried by the same chromosome. The parts of the chromosomes that determine the traits are called genes. This productive model was used to interpret the results of many breeding experiments (with fruit flies, molds, etc.), including those with mutated strains. Using the ideas of linked traits and mutations, geneticists were able to map out the assignments of various genes to their chromosomes. But this model still treats the genes themselves as unknown determiners of hereditary traits. There is no conceptualization of the nature or structure of the genes in the model.

The next refinement is to identify the gene as a biochemical agent; in other words, the genes express the hereditary characteristics by specifying the structures of proteins created by a cell. The proteins known as enzymes, which are responsible for much of the biochemical activity of life, are especially important in this model of gene action. The idea of genes as biochemical agents was based on a great deal of information from genetics experiments, combined with a growing understanding of the biochemical processes occurring in living cells. But the details of what the genes are made of, what the genes look like (their structure), and how the genes are able to specify the manufacture of proteins were totally unknown at the time this model (gene as biochemical agent) was devised.

To answer these questions about genes required the use of new techniques for determining the structure of large and complicated molecules. One important discovery, made with such techniques, was that nucleic acids (e.g., DNA) don't have any predetermined order in the sequence of bases that make up the molecule; the bases can be in any order. Because of this fact, the DNA molecule can encode information (more particularly, genetic information) in its sequence of bases. The DNA molecule was identified as the fundamental component of the genes, and the problems then became these: What is the spatial structure of the DNA molecule? How is the encoded information used to create proteins? What is the code? The structure of DNA was soon determined to be the famous double helix, and the details of its action (using RNA as an intermediary) were worked out over a number of years. We now have a highly detailed model of how hereditary information is transmitted in living organisms. The last refinement of the model was the discovery that some genes do not carry structural information about proteins. Instead, these genes control the actions of the structural genes, telling them when to turn on and turn off. In this model, the interpretation of the coded sequence of bases on the DNA molecule becomes rather more complicated. Molecular biologists now have a vastly greater amount of detail worked out, however, and this model is the currently accepted picture of hereditary transmission of traits. Notice that each of the successively improved models encompasses, and doesn't contradict, the model that precedes it.

Game Theory

Discussion of the social sciences is generally outside the scope of this book. I'll make an exception here, however, because models are very useful in the social sciences just as in the natural sciences. Some social science models are mathematical, especially in economics, while others are more conceptual. One of our major points so far has been that models are simplified (and sometimes overly simplified) representations of reality. It's probably fair to say that mathematical models in the social sciences need to make extreme simplifying approximations. The question then becomes this: What interesting and valid results do we get from such models? In other words, when are the approximations good enough? This question belongs at the forefront of social science research, and we can't really address it well here.

Instead, we'll look at one simple model of social interactions, and at the assumptions this model makes. The social interactions we'll consider are conflicts, and the model is the mathematical theory of games. Conflicts arise in a variety of circumstances. Political conflicts occur between groups in a society, for example, or between nations. Military conflicts can also occur between nations. In economics, conflicts between different companies or between management and labor might occur. The common element in all of these varied situations can be modeled as two or more participants engaged in a contest or struggle (of some sort) that has winners and losers (note the simplifying assumptions that we have already made). Now, a key point in most real conflicts is that each side will devise a strategy in order to win; but whether this strategy works depends on the strategy of the opponent, which is unknown. This element of uncertainty about the opponent's actions is the issue that game theory tries to address.

Game theory was invented in 1928 by the mathematician John von Neumann. He started by considering a very simple example: a game of matching pennies in which two players lay down pennies either heads up or heads down. If they match, one player wins (and gets the pennies). If they don't match, the other player wins. Now it's very obvious that you can't make a choice that guarantees you'll win; whether your choice wins depends on the choice your opponent makes, which you can't know. How, then, can you devise a strategy? Well, you can't devise a strategy that assures you'll win more than you lose. But, you can assure that you won't lose more than half the time (i.e., you'll break even in the long run). If you impose any kind of pattern on your play, you run the risk that your opponent will discern the pattern and outguess you (you'll lose more than you win). But if you lay down half heads and half tails at random, probability dictates that you won't lose more than half the time. Since this kind of play minimizes the losses of both players, von Neumann considered this to be a stable solution of the problem. This game, and our analysis, seem trivial because the game is so simple. There are only two possible moves, and winner takes all. Von Neumann's brilliant insight was to apply the same kind of reasoning to more complicated games. He considered games where there are many possible choices, and where each choice leads to different amounts of gain or loss for each player. The mathematical analysis then becomes decidedly nontrivial. Remarkably, von Neumann was able to show that for any zero-sum game (i.e., one player's loss equals the other player's gain) with two players and some random choices, there is always a strategy that minimizes the losses of both players. Given the particular rules of a game, we can mathematically determine that strategy.

In this analysis, we have made two major simplifying assumptions. We've restricted the game to just two players, and we have assumed that it's a zero-sum game. Game theorists have looked at multiplayer games in great detail. An interesting new feature is the possibility of coalitions forming and breaking up. We can then ask which of the possible coalitions are most advantageous to the players. The possibility of partners negotiating how to divide the winnings can also be built into the model, and the solution of the game then must include the optimum manner to do so. If the game is not zero-sum, then far more complex (and fascinating) possibilities arise. A famous example of a non-zero-sum game is known as the Prisoner's Dilemma. The players are two prisoners who have been arrested for committing a crime together; they are kept separate and can't talk to each other. If neither confesses, both go free. If both confess, both go to jail with a light sentence. If one confesses and the other does not, the confessor is paid off and the partner gets a heavy jail sentence. The game has a number of ethical and psychological dimensions that are worth considering. From a game theoretic point of view, though, the main interest is this: If each player employs a strategy to maximize personal gain, then both players wind up worse off than they need to be. Unfortunately, no solution to this dilemma has yet been found (although there is an optimum strategy for many repetitions of the game, namely: always do what has just been done to you).

The usefulness of game theory in creating social science models is that it incorporates some of the elements of real conflicts (not knowing an opponent's strategy, random variations that we can't control). The novel contribution of game theory has been to make these "messy" characteristics of human affairs amenable to mathematical treatment. On the other hand, the simplifying approximations of the models are very severe. The rules and the payoffs in real conflicts are not usually well specified. Also, humans don't always act in the rational manner required by the models. Still, game theory has proven to be a fertile method in the social sciences. Applications to legislative voting patterns, military tactics, international diplomacy, business decision making, labor-management disputes, and anthropological studies of other cultures have been made. Game theory has even been applied in the natural sciences, in studies of animal behavior.

§3. IMPROVING MODELS

A central idea of this chapter is that a model is an approximate representation of the real situation, and we can often improve our model to get a better approximation. Sometimes this requires an actual conceptual change in the model. In this section, however, we'll look at two specialized techniques in which the basic concepts of the model remain the same. In one technique, the details of the model are improved by adding successively smaller corrections; in the other technique, the model is improved by tinkering with its parameters.

Systematic Correction

We can use several different methods to improve a mathematical model systematically with smaller and smaller corrections. By doing so, we can make the model as close to reality as we please (which is to say, we can make the predictions of the model match the experimental data to any desired level of precision) by including corrections that are small enough. One method is to represent the results of the model as a sum of several terms, each term becoming progressively smaller. A different method is to use iteration, which means reworking the model over and over again and getting closer to the correct result each time. A sum of terms, of the sort I just mentioned, is known in mathematics as a series. The idea is that a complicated model can be represented by such a sum, even though each term of the sum is simple. If we keep only the first term (which is the biggest) then we have only a crude approximation to reality. If we keep the second term as a correction to the first, we have a better approximation. Retaining the third term makes the approximation even better, and so on. Since each term is simple, the model stays fairly simple; since each term is smaller than the previous term, we get closer and closer to a correct version of the model.

This seems very abstract, but we can illustrate the basic idea by a simple process of long division in arithmetic. Suppose you divide 527 by 32. Do it out by hand. What do you get? Notice that the answer you get, based on the process you used to get it, can be written as

$$527/32 = 10 + 6 + 0.4 + 0.06 + 0.008 + \ldots,$$

which we normally just write as 16.46875 (this number can just as well be thought of as the sum I wrote down). If you decide that computing every term is too much work, you can stop at some point and keep fewer terms as an approximation to the correct answer. This approximation to the exact answer may be all you need. For example, if you are dividing $527 among 32 people, then 16.47 is close enough, because the terms you left out are fractions of a penny. If you really hate arithmetic and don't mind losing some money, you can retain only two terms to get the crude approximation 16. The analogy of these procedures with successive approximation in models by adding ever-smaller correction terms is very close to the mark.

A scientific example of this method is the perturbation series technique used to find planetary orbits. Suppose we want to calculate the orbit of Saturn. A first approximation (which is already very good, incidentally) is to include only the gravitational influence of the sun. But the planet Jupiter is also pretty big (though small compared to the sun) and near Saturn, so a second approximation would be to include the gravitational influence of Jupiter on Saturn as a correction to the sun. In other words, we first ignore Jupiter and calculate an orbit based on the sun; then we include Jupiter as a perturbation on this orbit and calculate the small correction that this entails. We could extend this method to include the gravity of any other planet, or else we could leave other planets out because their corrections are too small to worry about (i.e., have no observable consequence). An extremely interesting historical episode involving perturbation theory is related in chapter 4.

The second method I wish to discuss briefly is the iterative method. In this case, all the terms of the model are large. We therefore can't include some terms as small corrections on the rest, but we may still be able to use successive iteration (repeating the same calculation over and over). We start by making some initial guess about the values of all the terms, and use this guess as a starting point to solve the model. The first essential requirement in this technique is that we have some recipe with which to calculate results based on our guess. The second requirement is that we also have a way to use these results in finding a new and corrected set of terms in the model. We then use our new values of the terms as a new starting point for another iteration, and we go on to calculate yet another new set of terms. If we are on the right track, these corrected terms are closer to the previous starting point in this second iteration than in the first iteration. We can continue iterating this way until the input terms and the output terms are the same, a condition known as self-consistency, which indicates that we have gotten our model right. As you might guess, this method has been made more practical by the tremendous increases in computing power in recent decades.

Parameter Variation

We have already seen in §2 that models often have parameters, numbers that represent important quantities in the model. Sometimes, parameters are measured independently (as in our traffic model). Parameters are also sometimes computed using some independent method that isn't part of the model itself. But sometimes, parameters are varied or adjusted until the model predictions match some data (as in the drug uptake model). We say that the parameters of the model are "fit" to the data. Parameter fitting can be a valuable technique, but it can also be a misleading technique. Let's look at a simple example to see how parameter fitting works in practice. Suppose we are trying to model how water consumption depends on population in a city. A good first guess is that this dependence is approximately linear, that is, the amount of water used is proportional to the number of people who live there (see chapter 19). So, we make a linear model to approximate this relationship. Our model has only one parameter, namely, the proportionality constant. To explore the issue further, we might make a plot of water consumption versus population, obtaining data from historical records. Now, we wouldn't expect every data point to lie exactly on a perfect straight line. But if our model is a good one, we expect our set of scattered points to cluster about a straight line. Parameter fitting in this case consists of finding the "best" line through all these points, the line for which the greatest number of points lie as close to the line as possible. The slope of this best-fit line is the desired value of the parameter in our model. (The name of the technical procedure used to find slopes in this way is a "least-squares fit," which is available on many commercial software packages.)

As I said, there are both good points and pitfalls to parameter fitting. If a model is essentially correct, parameter fitting gives us a way to extract valuable numerical information from the data. The pitfalls are related to the use of parameter fitting as a test of *whether* a model is correct. Perhaps, in our example, the dependence is not linear for some reason (e.g., shifting demographics). If the departure from linearity is small and the scatter is large, we may not notice. We would then confidently announce that our model (which is incorrect) must be right because we got a good fit to the data with it. The problem is minor in this case, because we'd soon notice if the model wasn't at least approximately right. The major problems arise in more complicated models, which have many parameters. By varying all of the parameters at once, it's often possible to fit a limited amount of data with a model that isn't even close to being right; the many adjustable parameters merely give the model enough flexibility to fit many different sets of data. We can't always conclude that a model is correct solely because it matches observations, if adjustable parameters have been used; the fitting procedure might simply force the model to match the observations. An ancient example of this problem is Ptolemy's geocentric model of the planetary motions (see chapter 5). The properties of each epicycle can be adjusted independently to get the best fit to the observations of the planetary positions. This geocentric model had a large number of parameters that could be varied to fit a relatively small amount of data. The result was a successful model that was, of course, thoroughly wrong in concept.

The dangers of parameter fitting have occasionally shown up in my own research work. Some colleagues and I once used a model potential with two adjustable parameters that we fit to five measured atom-surface binding energies. We obtained a very good fit, i.e., binding energies computed using this potential matched the measured energies quite well. However, we realized (on rather general theoretical grounds) that one of our parameters was only affected by extremely small binding energies, which our experiments could not measure. Our good fit did not ensure that our results were meaningful. The value of a parameterized model depends on our goals. If our only goal is to have a model that can serve as a tool in applications (regardless of its conceptual merits), then a parameterized model that fits all of the data of interest might be fine. If our goal is to achieve understanding, however, then we demand more from our parameterized models. For example, in a model with parameters representing well-defined quantities of interest, the best-fit values of these parameters can be compared with independently calculated values for these same quantities. If the calculations agree well with the numerical values of the parameters, we have good evidence that our ideas are on the right track. Using methods like this, even schematic models can give us insights about nature.

§4. The Physicist and the Horse

The use of approximations and models is so integrated into the fabric of science that it has even become the basis for a joke. I first heard this story many years ago, and since then I've come across several versions. It goes something like this: An organized crime syndicate has decided to ask several people how to predict the winners of horse races (the ones they are unable to fix, I suppose). The first person they ask is a psychic. The psychic gazes into a crystal ball and predicts the winner of a race. The horse loses. I'll leave the fate of the psychic to your imagination. The syndicate decides to ask a computer programmer next. The computer programmer writes a program to predict the winners of horse races, enters in all relevant information, runs the program, and announces the winner of the next race. The horse loses. The computer programmer shares the fate of the psychic. Finally, the syndicate decides to ask a physicist. The physicist tells them to come back in a few weeks and starts thinking about the problem. The syndicate comes back as scheduled, and the physicist says, "I'm not finished yet, come back in another week." They come back in another week, and the physicist says "All right, I'm ready now. But I won't just tell you the answer, I have to explain my method to you." So the physicist leads them all to a blackboard, sits them down, steps up to the board with a piece of chalk, draws a large circle on the board, and says: "First, assume that the horse is a sphere. . . ."

For Further Reading

"Theory of Games," by S. Vadja, in *The World of Mathematics*, edited by James R. Newman, Simon and Schuster, 1956.
The Nature of Scientific Thought, by Marshall Walker, Prentice-Hall, 1963.
Ideas of Science, by B. Dixon and G. Holister, Basil Blackwell, 1984.
"Maria Goeppert Mayer: Atoms, Molecules, and Nuclear Shells," by Karen E. Johnson, *Physics Today*, vol. 39, no. 9, p. 44, September 1986.

"A History of the Science and Technology Behind Gene Mapping and Sequencing," by Horace F. Judson, in *The Code of Codes*, edited by D. J. Kevles and L. Hood, Harvard University Press, 1992.

Game Theory and Strategy, by Philip D. Straffin, Mathematical Association of America, 1993.

EPIGRAPH REFERENCE: Kip S. Thorne, *Black Holes and Time Warps*, W. W. Norton, 1994, p. 191.

Chapter 7

THINKING STRAIGHT: EVIDENCE, REASON,

AND CRITICAL EVALUATION

> The number of persons who have a rational basis for their
> belief is probably infinitesimal; for illegitimate influences not
> only determine the convictions of those who do not examine,
> but usually give a dominating bias to the reasonings
> of those who do.
> *(William E. H. Lecky)*

> Crime is common. Logic is rare.
> *(Sherlock Holmes)*

A WELL-CONSTRUCTED scientific argument, defending a scientific conclusion, generally rests upon two foundations: reliable empirical evidence and sound logical reasoning. Of course, there have been scientific arguments that weren't based on good evidence and reasoning, but these shoddy arguments (and the conclusions based on them) generally don't withstand the test of time. The point isn't that we always have proper evidence and reasoning in the sciences, the point is that we always *should* have these things. This point is not trivial. In some areas of human thought, conclusions may quite properly not be based on logic and evidence. In some political discourse, for example, we might legitimately based our conclusions on a shared set of values and traditions instead of rational analysis (elected government versus divine right of kings, for instance). Although such considerations sometimes enter into scientific thinking (see chapter 11), science still provides an excellent example of reasoned discourse. As such, we can at least use science as a starting point for a discussion of valid argumentation across the board.

The subject matter of this chapter, as you see, extends far outside the boundaries of science. We are concerned here with methods of clear thinking and critical analysis that are relevant to any issues. Where the methods and thought processes typical of the sciences are applicable, these methods are invaluable. When issues turn on differences in values, faith, cultural background, and so on, then we are obligated to isolate these differences and identify them clearly. Differing values are no excuse for bad logic and lack of evidence. Valid argumentation in the murky and ambiguous issues of human affairs is not essentially different from valid argumentation in the esoteric realms of science; it's just more difficult.

§1. GOOD ARGUMENTS

Deductive Reasoning

The purpose of deductive logic is to find relationships between statements (called premises and conclusions) that guarantee the truth of the conclusions *if* the premises are true. On the one hand, this means that we must be very careful to scrutinize the premises carefully. False premises can lead to false conclusions even when the logic of the argument is valid. (Logicians use the word "argument" to mean a set of premises leading to a conclusion.) On the other hand, the ability to spot bad logic ensures that we don't get fooled into accepting false conclusions based on obviously true premises. In fact, if we can spot bad logic, then we don't even need to look very carefully at the premises; we already know the conclusions are suspect at best. A valid logical argument, based on well-supported premises, leads to a trustworthy conclusion. A few years ago, when I was exhorting one of my classes to look for logical flaws (along with undocumented assertions, errors of fact, internal contradictions, and so on), one of my students said "It's all a matter of opinion." But not everything is simply a matter of opinion. There are clear and well-defined rules worked out by logicians that can be used to analyze arguments. Recent work in logic is all symbolic, essentially reducing arguments to formulas and examining the conditions under which the formulas are valid. Although this work is interesting, I'll focus here on the older verbal tradition in logic, which goes back to Aristotle.

A central element of this tradition is the syllogism. A syllogism is a form of logical argument that consists of two premises and a conclusion. The premises might be general propositions, taken to be always universally true; particular propositions about a single object, person, event, and so on; or conditional propositions concerning the circumstances under which a statement is true. In a valid syllogism, the truth of the two premises ensures without fail that the conclusion is true. Logicians customarily illustrate the use of the syllogism with simple examples like this:

(1) All cats are cute. {major premise}

(2) Smokey is a cat. {minor premise}

(3) Therefore, Smokey is cute. {conclusion}

RP

Figure 5. If all cats are cute, then this individual cat must, by the inescapable logic of the syllogism, be cute.

(see Figure 5) It's quite obvious here that statements (1) and (2) cannot be true and yet have statement (3) false. People may have differing opinions about statement (1), and statement (2) is a simple matter of fact, either right or wrong. But if we accept both of these statements, the truth of statement (3) inevitably follows. Syllogisms like this one, which contain no conditional propositions, are called categorical syllogisms. The syllogism is a useful tool in constructing an argument. Once you have established a valid syllogism, you can then concentrate on establishing the validity of your premises. Perhaps more importantly, in analyzing someone else's argument, you can disentangle the logic from the premises. After you isolate the syllogistic form of the logic, you may see clearly that the reasoning is invalid. In that case, you don't have to worry about analyzing the premises (often a difficult and problematic task, leading to no unambiguous result). The veracity of the premises doesn't matter if the logic is invalid; the argument is no good anyway.

Another useful point concerning syllogisms is that the logic can be reversed. Once a valid syllogism has been established, one of the rules of logic tells us the following: If the conclusion is known to be false, then at least one of the premises *must* be false. Apply this reasoning to our example. If Smokey is not cute, then either Smokey isn't a cat or else not all cats are cute. I should also mention that there are some conclusions which cannot be drawn from syllogisms. For example, the truth of the conclusion doesn't tell us anything about the truth or falseness of the premises. Similarly, the falseness of the premises tells us nothing about the truth or falseness of the conclusion. We'll explore these rules further in our discussion of bad logic in §3. Finally, let's look briefly at another important part of deductive logic, the conditional proposition. One type of conditional proposition tells us something about the truth of statements that are paired together (example: either 1+1=2 or my understanding of arithmetic is wrong). A second important type is the hypothetical proposition, which takes the form of an "if . . . then" assertion (example: if the Orioles win one more game, then they will be in the playoffs). These conditional propositions can also be used as premises in a syllogism.

Inductive Reasoning

In deductive logic, our conclusions are based on a set of premises, and the truth of the premises implies the truth of the conclusions. The advantage of deductive logic is that we have the certainty of truth in those cases where the method can be used. The disadvantage of deductive logic is that we seldom have any well-defined general premises that we know are true. Instead, we can use a different form of reasoning, called inductive logic. The method of inductive logic is to use the truth of many particular statements to make a generalization, which is our conclusion. If every cat I've ever seen is cute, then I conclude based on this experience that all cats are cute. Obviously, the disadvantage of reasoning by induction (as opposed to deduction) is that my conclusions are less certain. I might run into an ugly cat tomorrow.

The use of inductive reasoning at some point is almost unavoidable. The difficulty is how to assess the validity of an inductive argument. Since a proof by induction can never be absolutely certain, how can we judge the quality of any conclusions drawn? Philosophers have put a lot of effort into deciding how to make such judgments; but for analyzing typical arguments, we can get a lot of use just from common sense. If a generalization is based on only one or two examples, then the conclusion is basically worthless. If the generalization is based on thousands of well-controlled and highly documented cases, then we can (at least provisionally) accept a conclusion in this case. In between these extremes, we'll accord conclusions the respect they deserve based on the amount of inductive evidence presented.

Evidence

As we've seen, a valid deductive argument only results in a genuinely valid conclusion if the premises are true. An inductive argument is likewise only meaningful if reliable particular cases are presented. In both of these methods, the quality of the argument rests as much on the evidence offered in support of the argument as it does on the logic. Accordingly, we must be able to evaluate the quality of this evidence in order to analyze the validity of the argument. How can we evaluate evidence? One of the first questions to ask is whether any documentation has been presented to back up the contentions being made. Why should we believe a claimed statement? What source, reference, or authority has warranted this claim? Where does a particular fact come from? Is the source of this fact reliable or not? Politicians and editorialists often feel no need to provide any documentation at all for claimed facts. When a source is given, it's often of doubtful credibility. Beyond asking for documentation, we can also make our own evaluations of how plausible a claim is. This involves looking for internal contradictions, violations of intuition and common sense, contradictions of other facts we know to be true, numerical estimates that don't make sense, and so on. Some of these methods are typically used in the sciences (§2), but critical evaluation of evidence is always a necessary part of a proper analysis (§3).

§2. THE CONTRIBUTIONS OF SCIENCE

What does critical analysis of argumentation have to do with science? One simple answer to this question is that science routinely employs this very kind of critical analysis all the time. Science is one of the few human endeavors in which we sometimes have the luxury of starting with a general premise, deductively working out the results of this premise, and comparing these results to evidence of extremely high quality. Examining this process at work in the relatively tidy and uncomplicated problems of science gives us a sense of how to proceed in the more difficult realms of politics, economics, social issues, and so on. The skeptical attitude characteristic of the sciences also fosters a spirit of critical analysis across a wide range of issues. Beyond these general considerations, however, science offers several techniques and modes of thinking that are not commonly found in other fields. In the rest of this section, we'll survey some of these scientific thinking practices.

Using Basic Knowledge

Although science continues to progress as new results are discovered and old theories are modified, we have a certain amount of core knowledge and experience in science that is not likely to change radically or quickly. Unlike the trends and fads of political ideologies and public opinion, we can count on this basic scientific knowledge to be correct. When such knowledge is relevant to the premises of an argument, then we can evaluate these premises without knowing every factual detail. An example is the conservation of energy law (see chapter 17). There are no known exceptions to this principle, and none are expected. If the success of some public policy initiative depends on the creation of energy from nothing, then we can reject this policy without further consideration. You don't need an extensive stock of scientific knowledge to apply this kind of reasoning. For example, it's clear on very fundamental grounds that the earth is finite. Anyone who argues that the earth's resources are limitless must then be wrong. Any argument that depends on limitless resources as a premise is likewise wrong. Another example of basic knowledge is the form and properties of exponential growth (chapter 20). Whether the subject of debate is population growth, economics, resource use, finance, or energy policy, any exponentially changing quantity shares the same properties and characteristics; and these properties are not subject to contrary opinion, they are simply matters of arithmetic. If the premise of an argument demands a contradiction of these known properties, then you may safely conclude that this premise is wrong. Notice that both of these examples provide us with constraints on truth, giving information about what must be wrong instead of telling us that something is right.

Probabilistic Thinking

At one end of the spectrum, these few rock-hard certainties are useful when we can apply them, but unfortunately this is seldom. In other words, we often find ourselves trying to arrive at conclusions in the absence of the information we need. So at the other end of the spectrum, we can employ a completely different way of thinking that scientists also find useful, namely, thinking in terms of the probability that a statement is correct. If we don't (perhaps can't) know for sure whether a premise is right or wrong, we need to make our best guess as to how likely the premise is to be right or wrong. This style of thinking is alien to many people. "A statement is either right or it's wrong. How can there be anything in between?" Some people feel a need to choose sides, even in the absence of information, and become wedded to their position. More sophisticated people might still choose sides, but they will remain aware that their choice could well be incorrect and keep an open mind. In both cases, though, a position is staked out.

In the course of scientific work, we are often faced with situations we don't understand because of insufficient information. Suppose there are three alternative explanations for an experimental result. We don't want to choose one of them without good evidence. The ideal procedure might be to design further experimental tests to weed out the inferior ideas. But this procedure costs time and effort, so we rank the ideas based on the probability of correctness for each one and test the best idea first. Our probability estimates may not be very accurate, but at least they are better than choosing at random. This approach is a very natural way to think about a scientific problem. Since all scientific results are in some sense provisional, thinking in terms of probabilities gives us a way to make distinctions between ideas that are speculative, those that are well founded, and those that are quite certain. The rock-hard certainties I mentioned before are the concepts with such a high probability of being correct that we can, for pragmatic purposes, assume they are true. A few things in everyday life fall into this category, like the inevitability of death and taxes. Most real-life issues, however, are much more complex and ambiguous than scientific questions. Does it make sense to attempt estimates for probabilities of correctness in these murkier cases? I believe that it does make sense to try, even if we don't have (and can't get) the additional information we need to verify and improve our estimates. The benefit I see in such probabilistic thinking is that we don't get tied to a position; we maintain a more fluid and flexible receptivity to new information and different viewpoints. On the other hand, probabilistic thinking allows us to make judgments rather than just give up because we don't have certainty on an issue; an application of this outlook to real-life public policy issues is discussed in chapter 10.

Hidden Assumptions

The conclusion of an argument is based on the premises of the argument. In verbal rhetoric, however, the premises are not always stated clearly. Sometimes the premises are implied or taken for granted. A syllogism containing an unstated but clearly implied premise is called an enthymeme ("Sacco is evil because he is an anarchist" implies the unstated major premise "all anarchists are evil" and also states the conclusion first rather than last). But even when the premises appear to be stated fully, there might be some further hidden premises assumed, either in addition to those that are stated or else underlying those that are stated. In the analysis of an argument, either scientific or nonscientific, it's always important to look for hidden assumptions and to evaluate the validity of those assumptions. Let's look at a few scientific examples. If a chemical reaction needs to be done in the absence of oxygen, a chemist might do the experiment in a container with the air pumped out. The unstated underlying assumption here is that the tiny amount of oxygen left over (about 0.01 percent of the amount found in air) is not enough to affect the experiment—probably a good assumption, but not necessarily. A famous example of an unstated assumption (in physics) was that the measured velocity of light would depend on the velocity of the person making the measurement. This "fact" was taken for granted until Albert Einstein stated explicitly that it was really just an assumption, and an incorrect assumption too. In biology, the classification of organisms into animals and plants carries with it the underlying assumption that all organisms must fit into one category or the other (this assumption has also been challenged by more recently discovered organisms). For a long time, an unstated assumption of medical research was that studies having only adult males for subjects produce results applicable without modification to the rest of the population.

Now let's apply the technique to some political issues. In the debate over gun control legislation, both proponents and opponents make unstated assumptions about the relationship between the incidence of violent crime and the easy legal availability of guns. In debates over the desirability of environmental regulations, there is often a hidden assumption embedded in the arguments, namely that such regulations are a drain on the economy leading to loss of jobs, and so forth. Politicians proposing large tax cuts often employ the underlying assumption that no relationship exists between the tax revenues collected by the government and the desired services provided by the government. A hidden assumption may or may not be correct, but until we bring it out into the open by making an explicit statement of the assumption, we can't engage in an analysis of its correctness.

Evaluating Causality

Although philosophers differ over some of the finer points concerning causality, we do have some pragmatic criteria for establishing causality in both science and logic. These criteria are just as applicable in everyday life and public affairs as in the sciences. Many people find the subject confusing, and invalid claims of cause/effect relationships are pretty common. One mistake is so common that it even has a Latin name: *post hoc, ergo propter hoc*. A literal translation is "after this, therefore because of this." When one event follows another, you might assume that the first was the cause of the second. After Jimmy Carter was elected President, the country suffered a period of high inflation. Can we conclude that Carter's policies caused the inflation? (His political enemies certainly made this claim.) But his term had been preceded by many years of high military and domestic spending; during his term, oil cartels had artificially driven up the price of energy. Both of these conditions, having little to do with Carter's policies, are more plausible reasons for the cause of inflation at that time (actually, of course, a complicated detailed analysis is needed here). On a less grand scale, consider your car; suppose it needs some repairs to the engine, then a new muffler, then a new battery, and finally some new tires. Does each of these problems cause the problem that follows? I doubt it. This example, in fact, illustrates an alternative explanation that makes more sense than "*A* causes *B*." A more likely situation is that the car is old, and the old age of the car is responsible for all of the other problems. "*C* causes both *A* and *B*" instead of "*A* causes *B*" is often a possibility worth considering.

Fallacious reasoning about causality is ubiquitous in our political and economic discussions. Any time people see correlations between events, trends, or quantitative measures, they have a strong desire to assume a causal link. But there may be many different causes contributing to a single effect. Or the two correlated things may both be caused by something else that you haven't identified yet. Or you may not have a cause/effect relationship at all in some cases; for example, two events might be related by a feedback loop (see chapter 21). And of course, the correlation you see may be nothing more than a coincidence. The welfare state has caused an increase in poverty; guns on the streets have caused an increase in violent crime; environmental regulations have caused a decrease in productivity; sexual immorality has caused the AIDS epidemic; television has caused a declining attention span in our youth; and so on. Claims like these, which vary greatly in plausibility, are made all the time. Very few, if any, are actually valid claims.

How can we rigorously demonstrate a causal link? Doing so turns out to be very difficult. We would first need to demonstrate that the cause *must* have been present for the effect to occur. In addition, we would need to demonstrate that the effect will *always* occur when the cause is present. In the language of logic, we say that the cause must be both a necessary and a sufficient condition for the effect. In science, this can sometimes be accomplished by a detailed series of carefully controlled experiments. Much of the controversy and confusion arising from biomedical studies results from the ability of such studies to draw conclusions that are highly suggestive of causal links, and their inability (due to monetary and/or ethical constraints) to rigorously prove causality. Since the majority of complicated situations have multiple partial causes, we still customarily use the word "cause" even when the cause is neither necessary nor sufficient. In these cases, we must settle for a statistical inference of causality, requiring a large random sample. For a one-time historical event, a rigorous demonstration of causality is virtually impossible. The best we can do is to make a detailed analysis that accounts for as many known possible causal factors as we can think of and assess the role of each one in bringing about the effect. We will undoubtedly not be able to prove causality, but we may well be able to make a convincing case (probability again).

Models

Another important ingredient in establishing causality is having a causal model. In other words, we should have some reasonable way to understand how and why *A* causes *B*. A good model by which to understand the claimed causal link contributes to making the claim more believable. The causal link between tobacco smoking and lung cancer, for example, is based in part on statistical evidence and in part on physiological models of the carcinogenic activity of the tars. Establishing causality is only one of the many important uses of models in thinking, both inside and outside the sciences. The topic of models, modeling, and successive approximation is so important that I have given it an entire chapter (6) of its own. One worthwhile point concerning models and the evaluation of arguments is this: Arguments are often based on analogies, and the validity of the argument then turns on how good the analogy is. An analogy is a comparison between dissimilar ideas, events, processes, and so forth; the comparison might be very apt, but it might also be wildly inappropriate. How can we judge whether an analogy is good or not? Since models are also basically a kind of analogy, familiarity with models (and how to assess them) can help us to evaluate critically a claim based on analogy.

Quantitative Thinking

Many of the arguments that we must evaluate are totally verbal, but sometimes an argument rests at least partly on some numerical claims. These numerical arguments require their own special techniques of critical analysis. How accurate and how precise are the claimed numbers? Do the numbers make any sense in terms of other things you know? Can you make an order-of-magnitude estimate of your own against which to check the claims? Are the numbers as large or as small as claimed when you compare them to something relevant (e.g., a percent change)? Once again, science offers us a variety of useful ways to think about these quantitative issues, and the importance of the material warrants a separate chapter (8).

§3. Bad Arguments

An argument can be bad in a variety of different ways. A bad argument simply means an argument that has an untrustworthy conclusion, as opposed to being invalid in a formal sense. Logicians call this "material" validity. A materially invalid (bad) argument might be defective on several different grounds. For example, the argument might simply be logically flawed as we discussed in §1. But, the argument might instead be logically valid and have a false conclusion because some premises are false. For example:

(1) All cows are reptiles.

(2) All reptiles can fly.

(3) Therefore, all cows can fly.

This ridiculous example is a logically valid syllogism, in a formal sense, even though not a single statement in it is true. Another type of invalid argument is one in which the statements are not necessarily false, but the statements are so ambiguous that they don't have any well-defined meaning. Many rhetorical devices are also used by writers and speakers to convince without valid logic or evidence. A number of these fallacious arguments have been categorized and named. The typical political speech, newspaper editorial, or magazine opinion piece is far more likely to contain fallacious rhetorical tricks than any actual attempt at valid argumentation. We'll examine a few of the more common types of fallacy (straw man, false dilemma, ad hominem, begging the question, and slippery slope) later. Next, though, we'll examine some of the mistakes that are often made in the logical form of arguments, and then take a critical look at the validity of evidence used in arguments.

Invalid Logic

Let's start with a look at arguments that do attempt some logical structure, but that have a logical flaw. After starting with a true (or at least plausible) argument, one mistake commonly made is to assume the truth of its converse (i.e., reversing the conclusion and premise). For example, you might read a well-documented and convincing essay arguing that overly high taxation rates impair the productivity of an economy. At the end, the writer demonstrates that our economic productivity is low, and from this concludes that we have an overly high taxation rate. Note the form of the logic: "If taxes are too high, then productivity is low; productivity is in fact low, therefore taxes are too high." Even if you accept the first statement, the second statement need not be true. A valid argument doesn't imply its converse (after all, productivity may be low for a different reason). If the flaw in the logic is not apparent, consider a more obvious example: "If I am a corporate executive, then I make a good salary; I do make a good salary, therefore I must be a corporate executive." But of course it's not so. Maybe I'm an M.D. or a basketball star.

The inverse of an argument is another example of invalid logic. Once again we start with a valid argument, so that some premise really does imply a conclusion. We then show that the premise is wrong. Because the premise is false, we decide that the conclusion has also been proven false. For example, consider this fallacious bit of logic that was used often a few decades ago: "Citizens who support the country's war effort are patriotic citizens. Therefore, those who don't support the war are not patriotic." The truth of the first statement in no way implies the truth of the second. (After all, very patriotic people may well believe a war is not in the best interests of their country.) If emotions are running high over an issue like this, the illogic of the argument may well be overlooked. The incorrect logic becomes more obvious in this example: "All dimes are coins. This is not a dime, so it can't be a coin." But it may be a nickel.

Our last example of commonly used invalid logic concerns the incorrect use of what logicians refer to technically as distribution. Distribution means making a statement about all members of a class. When we say that all cats are cute, the term "cats" is distributed because we have made an assertion about every single cat. The term "cute" is not distributed in our statement because we have not made any assertions about all possible cute creatures (or any other cute things, for that matter). You can see the possibilities for foggy thinking here.

For example, we can make a simple invalid syllogism:

(1) All cats are cute.

(2) All dogs are cute.

(3) Therefore, all cats are dogs.

This example illustrates the fallacy of the undistributed middle term. The middle term of a syllogism is the term that appears in both of the premises, the term that relates them to each other in some way (e.g., "cat" is the middle term of our original syllogism example back in §1). A rule of logic states that the middle term must be distributed in at least one of the premises in order for any valid conclusion to be drawn. Our example makes the mistake seem silly and trivial, but people's lives have been destroyed on the basis of this logical fallacy. "Mr. Jones has admitted that he is member of the Free Oppressed Peoples League (FOPL). The FOPL has been proven to be a Communist front organization. Jones is obviously a Communist and should be fired from his government job." This kind of reasoning was all too common a half century ago. The middle term of this syllogism is the FOPL. FOPL is undistributed in the first premise because it's the predicate of the sentence. The tricky reasoning is in the second premise, the logic of which should be restated as "some members of the FOPL are Communists." (No more than that can be validly inferred from the FOPL's status as a Communist front.) Restated in this way, we see more clearly that the middle term is undistributed ("some" not "all") and that no valid conclusions can be made concerning Mr. Jones. There are a number of other ways in which invalid arguments can be made by using distribution incorrectly. Without going into a lot of technical details, we can summarize the basic gist of the rules fairly simply: You can't draw a conclusion that is stronger than the premises on which the conclusion is based.

Invalid Evidence

In typical disputes over everyday matters and public affairs, inadequate evidence is at least as big a problem as flawed logic. There are several different varieties of invalid argument based on evidence; most of them can be found in virtually any edition of any newspaper. The simplest example of invalid evidence is the old-fashioned unadorned lie. People who are opportunists or extreme ideologues will say virtually anything, with little regard for whether it's false or true. In some cases, we have no trouble discerning lies; few people pay any attention to the claims made by spokespersons for the tobacco industry. In other cases, though, we would need some kind of independent reliable information in order to detect lies. A plausible statement from an apparently reputable source is usually taken at face value. Our only recourse in these situations is to have as many different information sources as possible, and to accept statements only on a tentative basis (in the spirit of probabilistic thinking discussed in §2). We do have a few strategies with which to defend ourselves against lies. A claim which *sounds* plausible might become more dubious upon further critical evaluation (an example is given in chapter 8). We can also look for internal contradictions in a set of statements made by the same person. The reliability of the source is another place to check carefully and critically; many highly biased organizations give themselves neutral-sounding names and paid operatives now mount phony grassroots campaigns on various issues.

But people don't need to lie in order to mislead us. There is also the problem of suppressed evidence, also sometimes called an error of omission. We are told the truth, but only a part of the truth. A politician might say that she only has five thousand dollars in her bank account, a perfectly accurate statement, but one that fails to mention the two million dollars she has in stocks. A commercial advertisement might boast (truthfully) about the virtues of the ingredients in a product, but leave out the fact that all of the product's competitors contain the same ingredients. A variation of the suppressed evidence technique is the selective quotation, often called quoting out of context. As we all know, a person's actual thinking can be completely misrepresented by quoting a single sentence fragment from an hour-long speech. The quoted words are accurate (and the speaker can't deny saying them), but the position attributed to the speaker is totally distorted. Once again, we are faced with a difficulty. How do we know what we're *not* being told? We may be lucky enough to have the facts from another source, but we can't always depend on this. Our only alternative is to consider all claims critically and skeptically. Although we can't always know what evidence has been suppressed, it's often easy to hypothesize the kind of evidence that may well have been suppressed. In the end, we must make judgments and evaluations based on a variety of considerations (the consistency of a claim with other things we know; the reliability of the source; the plausibility of the claim; the plausibility of the hypothetical suppressed information). Some of the techniques outlined in §2 are helpful in this process. Obviously, the more information we have from a variety of different (and reasonably reliable) sources, the better able we are to spot cases of suppressed evidence.

At least two other criteria are useful in evaluating evidence. One is the source of the evidence. Often, no source at all is given; a fact seems to come from nowhere, or else a vague and undocumented source is given ("Many scientists agree that ———" or "Senior officials in the administration say that ———"). A variation of this problem is when a specific source is cited, but the worth of the source is questionable. I recently read a diatribe against vaccination that contained many quotations, facts, and statistics, each one carefully documented by citing a source; but the source was the same book in every single instance, a book written by somebody with no known qualifications and published by a publisher I've never heard of. Despite the copious documentation, I was less than impressed by the reliability of the evidence. The second criterion we can use is the likelihood that a writer (or source cited) really knows (or even *could* know) the fact being stated. Political parties, candidates, ideologues, and governments often tell you the intentions and motivations of their opponents; if you think about it, there's no one less likely to really know these things. You-are-there style journalism often tells you what someone was thinking or saying at a certain time, information not known by anyone with certainty. In cases like these, a little thought reveals that something stated as a fact is probably nothing more than a supposition. The factual-sounding form of the statement is there for rhetorical effect.

Statistics

Most of these points concerning evidence in general apply equally well to statistical evidence, but the use of statistics presents some extra opportunities to mislead. Statistical facts may be cited selectively, for example, which is another version of suppressed evidence. But statistical facts might be unreliable or misleading for more technical reasons, which we will now explore. The reliability of statistical results depends (among other things) on the size and quality of the sample used. Conclusions based on small samples tend to be quite unreliable. Some biomedical studies, which received a good deal of publicity, have actually been based on samples too small to tell us anything reliable. Even a large sample might yield misleading results. To tell us something unambiguously, a statistical sample must be chosen randomly. If you do a survey of people chosen from the membership of the Sierra Club, you probably won't get an accurate picture of American opinion on environmental issues. On a more subtle and realistic note, statistical studies may also be biased by a nonrandom response rate; in other words, the people who choose to send back a completed survey may hold opinions that don't accurately reflect those of the entire initial sample. Similarly, people who volunteer to take an experimental heart disease drug may well be people who already take extra measures (such as proper diet and exercise) to prevent heart disease.

Public opinion surveys are especially prone to biased results because the way in which the questions are worded has a large effect on the answers given. For this reason, you should view with suspicion any statistical claims about people's opinions. Many people will be in favor of helping prevent children from starving, while few people will support wasting more money on welfare cheats, despite the fact that both statements may refer to the identical policy change. Without knowing how the question was phrased, statistics concerning public opinion are as worthless as they are exact-sounding. Governmental statistical reports also suffer from various flaws. Crime rate statistics, for example, are based on crimes reported rather than crimes committed, which may not always be the same (rapes and crimes against poor ethnic minorities have often been underreported). Economic statistics are also dependent on the quality of the data reported to the government, which is quite possibly less than accurate in some cases.

Perhaps the major intentional use of statistics to mislead people employs accurate numbers that are then incorrectly interpreted, usually by leaving out some important point (suppressed evidence again). For example, a political party might claim that its budget proposal contains a 1.2 percent increase in money for some popular program. But if the inflation rate is 4.5 percent, this "increase" represents a substantial cut. If I want to know whether defense spending is increasing or decreasing, should I look at the dollars spent on defense or at the percent of the budget spent on defense? For a statistical comparison to mean anything, the quantities we compare must be appropriate. Unfortunately, we are rarely provided with all of the information we need in order to make such appropriate comparisons. Our only recourse is to question the information we *are* given and ask ourselves what we really want to know.

Rhetoric and Fallacy

There are many rhetorical devices that are intended to deceive a reader or listener with fallacious arguments. A variety of these techniques have been catalogued and named. Let's look at a few of the more common types.

A "straw man" argument is directed not at someone's actual position, but rather at a distorted version that was fabricated by the perpetrator of the straw man fallacy. This distorted version (the straw man) might be a weaker argument, with irrelevant evidence and poor logic substituted for the valid arguments that support the conclusion under attack. Alternatively, the straw man position might be a more radical version than the real position held by a person or group, and this radical distortion is more easily attacked (extreme positions rarely have much support in the general public). The straw man fallacy is a popular rhetorical trick used by ideologues, editorialists, and (almost universally) political campaigners.

Another often-used technique is the "ad hominem" argument. A literal translation of this Latin phrase is "to the man." In other words, instead of actually addressing the evidence and logic of a person's argument, those who commit the ad hominem fallacy attack the person herself. For example, suppose Senator Krupt has proposed campaign finance reform legislation. Opponents of this legislation, instead of saying why it will be ineffective or undesirable, merely attack the legislation by pointing out that Senator Krupt has often engaged in the sleazy campaign financing practices that this legislation will prevent. Possibly true, but definitely irrelevant. The effectiveness of the ad hominem argument in popular discourse is remarkable when so little thought is needed to reveal its fallaciousness; the validity of a position certainly doesn't depend on the virtues (or vices) of the person proposing that position.

The "false dilemma" is another piece of effective rhetoric that is logically invalid. In the false dilemma, two alternatives are proposed as the only possible positions that can be adopted. One of these is inherently weak and easily attacked. After demonstrating how poor this weaker position is, the argument concludes that the other alternative is correct. The logical flaw, of course, is that there may well be other alternatives besides those two, perhaps even a broad spectrum of possibilities that haven't been included for consideration. An oversimplified example of the false dilemma is the following: "We must engage in a massive military buildup because unilateral disarmament will surely lead to the destruction of our country." Other false dilemmas are "either good jobs or a livable environment" and "either traditional health care delivery or socialized medicine." These examples may sound silly when written out so starkly, but the same basic arguments might be quite plausible sounding (to the unwary) if they are dressed in enough rhetorical embellishment.

When an argument assumes the truth of the conclusion as part of the premises, this is known as begging the question (sometimes called circular reasoning). For example, suppose we argue as follows: "The government spends too much of our hard-earned money. Therefore, we need to cut back on government spending." We might be able to make a pretty good case for this conclusion, but we haven't made any kind of case at all here; we could just as easily have reversed the conclusion and the premise. We have merely begged the question.

Although there are many other categories of fallacious argument, we'll just consider one more, namely the "slippery slope." In this case, the position argued against is assumed to lead inevitably to some terrible result. This terrible result is then argued against, instead of the actual proposed position. Examples: "Banning the sale of machine guns is a bad idea because if we do that today, then tomorrow we'll be confiscating every gun from every citizen." "If we allow any logging in this old growth forest, then we'll soon be clearcutting the entire area." Unless some good evidence or convincing reasons are given to support the claim that one action really will lead to another, the slippery slope argument is a fallacy. A famous historical example of a slippery slope argument is the domino theory used during the controversy over the Vietnam war. Fallacious arguments can be difficult to spot because the stylistic form of an essay (or a speech) often disguises the logical structure. Conclusions and premises might be mixed together in any order, and some premises might simply be left out (unstated, either implied or hidden). Arguments can be based on analogy or appeals to emotion instead of logic, and the rhetorical use of loaded words (or images) can easily sway our opinions. Suppressed and/or fabricated evidence isn't always easy to detect, and our own prejudices can mislead us as much as clever rhetoric. But all these difficulties can be overcome to some extent by critically evaluating the arguments we encounter.

FOR FURTHER READING

Valid Thinking, by Philip Wheelwright, Odyssey Press, 1962.
The Art of Making Sense, by Lionel Ruby, J. B. Lippincott, 1968.
Logic and Contemporary Rhetoric, by Howard Kahane, Wadsworth Publishing Co., 1971.
Logic, by Wesley C. Salmon, Prentice-Hall, 1984.
The Informed Argument, by Robert K. Miller, Harcourt Brace Jovanovich, 1992.

A Beginner's Guide to Scientific Method, by Stephen S. Carey, Wadsworth Publishing Co., 1994.

EPIGRAPH REFERENCES: William E. H. Lecky, *History of the Rise and Influence of Rationalism in Europe*, Longmans, Green, and Co., 1910, p. xv. Sherlock Holmes, from "The Adventure of the Copper Beeches," in *The Complete Sherlock Holmes*, by Sir Arthur Conan Doyle, Doubleday, 1956, p. 317.

Chapter 8

THE NUMBERS GAME: USES OF

QUANTITATIVE REASONING

> When you can measure what you are speaking about
> and express it in numbers, you know something about
> it, and when you cannot measure it, when you cannot
> express it in numbers, your knowledge is of a meagre
> and unsatisfactory kind.
> *(Lord Kelvin)*

> Quantification sharpens the image of things seen by
> the mind's eye, both external phenomena and internal
> conceptions, but it is useless or worse until the right
> things are in the field of viewpoint.
> *(R. W. Gerard)*

§1. WHY NUMBERS?

THE WORD "quantitative" means measurable in numbers, as opposed to "qualitative," which refers to verbal description. Although not every aspect of science is quantitative, the sciences are certainly more quantitative than other intellectual pursuits like literature or philosophy. Scientific discourse is also more quantitative than typical everyday conversations. Why should this be so? What is gained by the process of reducing qualities to numbers, and what is lost? One major advantage of quantification is exactitude. Instead of saying that an elephant is heavy or that an atom is small, we can provide a number for the mass of the elephant or the size of the atom. (Of course the matter is a bit more complicated because no measured number is really exact, but instead is only as good as the measurement that we made to get it; see §2.) Our number may not be exact, but we have clearly gained in precision by switching from a word like "heavy" to a number like 1850 kg for the mass of our elephant.

We have actually gained something even more important than precision because the word "heavy" has no meaning by itself. An elephant is heavy compared to a hummingbird but is not heavy compared to Mt. Everest. Is an elephant heavy? Having a number for the mass of the elephant, we can now compare this number, quantitatively, to the masses of other things. We can say which is heavier, and we can say by how much. Our comparisons are unambiguous now, so we have gained in clarity as well as precision.

These advantages are found in everyday life as well as in science. Having a number for your bank account balance is probably better than feeling wealthy or poor. In scientific work, however, the matter becomes crucially important because a defining characteristic of science is that we compare our understanding of nature with the observation of nature. A qualitative agreement between the two might be possible for many different understandings, not all of them correct. Quantitative agreement between a predicted number and a measured number is much less likely to be the result of incorrect thinking. Physicists of the nineteenth century, for example, predicted that a hot hollow body with a small hole would emit infrared radiation out of the hole. Their theories also predicted, quantitatively, how much radiation should come out at each wavelength. At short wavelengths, the predictions disagreed with the experimental measurements of the radiation. This seemingly minor quantitative discrepancy was the beginning of the revolution in thought we now know as quantum theory, the technological fruits of which include lasers and microelectronic computer chips.

Example: The Area Needed for Solar Cells

More subtle and less obvious advantages also result from quantification and numerical work. It's possible to simplify complicated chains of reasoning, which would otherwise be difficult to carry out, by reducing them to a set of numerical computations. Consider the following example: Opponents of renewable energy sources sometimes claim that electrical energy from the sun is impractical because we would have to cover up vast areas of the country with solar cell panels to meet our needs. Is this true? It certainly sounds plausible enough. But how can we determine if the claim really makes sense or not? Verbal argumentation fails us here, and we need to think through the question numerically. To avoid worrying about the rate of energy use, let's just think about one day. We need to know how much energy is produced by a given area of solar cell panel during this day (i.e., we want the energy per unit area). We then need to know how much total energy is used by the country in a day. A little thought reveals that the total area of solar panels needed is the total energy used in the country divided by the energy produced per unit area. To obtain the numbers we need, we can start by looking up the yearly U.S. consumption of electrical energy (then divide by 365 days). Doing so, we find that the U.S. uses about 7 billion kilowatt-hours each day (the kilowatt-hour is the energy unit used by electrical utilities). Now, about 1 kilowatt-hour of energy is contained in bright sunlight shining for one hour on one square meter of ground. For want of a better estimate, let's take the number of sunny hours to be five, on the average. A typical solar cell has an efficiency of about 10%, that is, one-tenth of the energy in the light is converted to electricity. So, the energy per unit area is about 0.5 kilowatt-hour of electrical energy per square meter in a day. We now have both numbers we need. The arithmetic is easy: 7 billion total kilowatt-hours divided by 0.5 kilowatt-hour per square meter. We therefore conclude that we need about 14 billion square meters of solar panel. This sounds like a lot of area. But is it really a lot? Recall the relationship between area and distance (see Chapter 16). This area is actually only about 120,000 meters on a side, that is, a square that measures about 75 miles by 75 miles. For comparison, the U.S. is about 2000 miles by 3000 miles.

The main point here is that we could not have arrived at our conclusion (namely, that the claims of the solar power opponents are wildly incorrect) without using a numerical argument. Two other points are also worth making, though: First, our conclusion is not much affected by particular choices for unknown numbers (e.g., what would be the effect of assuming only one hour of sunlight per day instead of five?). Second, we have not shown that solar energy is practical (economic issues, raw materials availability, energy storage problems, etc., haven't been addressed); we have merely shown that one argument against solar energy is bogus.

Other Gains and Losses

If the phenomena we are trying to understand have been reduced to numbers, then the behavior of the phenomena can be represented by mathematical relationships among the numbers. Not only do the mathematical forms bring order and simplicity to complex phenomena, but manipulation of the mathematical relations can reveal new and previously unsuspected behaviors. We have now crossed the bridge from quantification to the role of mathematical thought itself in the sciences, so we'll not proceed further here. Instead, we will return to the question of whether anything is lost in the process of reducing the world to numbers. We certainly pay a price in the sense that we have lost both the raw sensory experience of what we study and any aesthetic dimensions it may possess. I prefer listening to the music generated from the quantitative information stored on a compact disc, rather than counting the etch pits on the disc. In this sense, many things outside the realm of science cannot be quantified and still retain their true meaning. A more interesting question is whether anything of genuine scientific interest is not amenable to quantification. This question may be more controversial, but it seems clear to me that the answer is yes. Taxonomy, for example, is the science of classifying plants and animals into an organized system; it's inherently descriptive and nonquantitative, yet central to science. But these larger issues are not our main concern here. Let's return to the more practical questions of how to utilize numbers effectively when they are appropriate to the problems under consideration.

§2. PRECISION

How Good Are Measured Numbers?

When you see a number, do you ever ask yourself: "How well do I know this number?" Most people probably don't (after all, a number is a number, right?). Let's consider an example we are all familiar with, such as the nutritional information on a box of cereal. Suppose our Wonder Flakes box tells us that one 8 oz serving supplies 15% of our daily requirement of carbohydrates. This sounds unambiguous enough, but think about it for a minute. Can that 15% possibly be the same number for a 250-lb person who does manual labor all day and a 140-lb person who sits at a desk all day? Presumably, 15% is some kind of average. Even granting that we have an average number, is our number necessarily exactly 15% or might it be 14.5%? How about 16.2%? Could it be as far off as 12%? Do you think that even 15.01% is too far off? Or do you think it surely must be exactly 15% or else they wouldn't have written 15%?

Actually, very few numbers are exact. The only examples I can think of are: purely mathematical numbers, quantities defined by convention (i.e., mutual agreement), and integers (I have exactly 10 fingers). Any other numbers that are the result of measurement are not exact. The proper interpretation of such a number is that it represents a *range* of values around the actual written value. How wide is this range? In the case of our Wonder Flakes box, we unfortunately have no idea. In fact, a number usually represents a range that we don't know. Keeping this point in mind is extremely important when evaluating the information that crosses your path each day. An unadorned number, with no information given about its precision, doesn't tell you as much as it pretends to. Knowing this, you can interpret numbers more realistically and thereby (paradoxically) get more genuine information from them. So, what does the 15% on the Wonder Flakes box really tell us? First of all, if the people who determined that number did a proper scientific job, then its precision is known and can be stated quantitatively (using technical methods described later in the section). But even if someone does know how precise that 15% is, we do not. We still need to form the best conclusions we can, given our limited information. To do this, we need to introduce the concept of an order of magnitude.

Orders of Magnitude

Since we don't know anything about the precision of this 15%, a prudent course of action is to assume the worst and take it to be very imprecise. Even so, the 15% is still telling us something important, because we know that it's in the general ballpark of about 10%, as opposed to either 1% (which would be a trivial amount) or 100% (which would be all the carbohydrates we need). You may object that 1% is not always trivial. And where did I get 1% from anyway, why not 2% or 1.5%? The answer to the first of these objections is that 1% *is* trivial in this context: nobody is going to eat 800 oz of cereal in a day. The answer to the second objection follows from the same reasoning; after all, nobody is going to eat 400 oz in a day either. Anything anywhere near 1% will be trivial, which is the major point here. The same logic applies to the 100%. We don't care much whether it's really 80% or 110%, because anything anywhere near 100% is almost all you need to eat. More importantly, anything near either 100% or 1% is obviously not right for Wonder Flakes. How do I know? Because both are inconsistent with the 15%, no matter how imprecise it is, if this number has any meaning at all.

Notice that 1%, 10%, and 100% are separated by factors of ten, and that they can be written as 0.01, 0.1, and 1.0. The closest power of ten to a number is called the order of magnitude of that number. For example, the order of magnitude of 0.15 is 0.1 and the order of magnitude of 889 is 1000. A convenient notation is to write numbers (after rounding off) as 10 to the appropriate power. For example, the order of magnitude of 114 would be written 10^2 and 0.0027 would be 10^{-3}. This method, called scientific notation, is very handy for expressing extremely large and small numbers. The method is also convenient for multiplying numbers because we only need to add the powers of ten together.

We now see that the only firm conclusion we can draw from the number 15% is that the order of magnitude of our carbohydrate needs supplied by this cereal is 0.1 (10^{-1}). This is dependable information because the actual number is highly unlikely to be an order of magnitude less or an order of magnitude more. The information is also worthwhile (after all, is there any reason you would ever want to know the nutritional content of your Wonder Flakes more precisely than its order of magnitude?). As a bonus, we have also automatically taken care of the fact that 15% must be an average because the carbohydrate needs of anyone will undoubtedly have this same order of magnitude. Sometimes, of course, you need more precision than this. I would not want my eyeglass prescription filled only to the nearest order of magnitude, for example. In a surprising number of cases, however, order of magnitude information turns out to be very useful. Let's illustrate this with two examples, one of purely scientific interest and another more concerned with public policy.

Consider the order of magnitude of the energy involved in chemical bonding. Using the energy unit known as the eV (abbreviation for electron-volt), the order of magnitude of chemical bonding energies turns out to be 1 eV. Compare this to the order of magnitude of the energy of light (you can think of light as consisting of discrete bundles of energy known as photons). A light photon also has an energy with order of magnitude 1 eV. So, the binding energies of chemical reactions and light photons have the same order of magnitude (although individual reactions of different chemicals have different energies, as do different colors of light). This similarity in orders of magnitude is interesting, because it allows several crucially important processes to occur. One example is photosynthesis in plants, the conversion of light energy into chemical energy stored in the body of the plant. Photosynthesis is obviously of great importance to us, ultimately providing all of our food. A second example is the action of the human eye's retina. Light falling on the retina causes chemical reactions that initiate the nerve impulses to the brain, resulting finally in what we perceive as visual images. The essential first step in both of these processes depends on a proper match between the order of magnitude of light energy and of chemical energy.

Let's now consider an example that is more concerned with practical matters of public policy. In the early 1980s, a program called the Strategic Defense Initiative (popularly known as Star Wars) was begun. The purpose of this program was to develop directed energy weapons such as lasers that could shoot down incoming nuclear bombs carried by intercontinental ballistic missiles before they could reach the United States. This program, costing tens of billions of dollars, was greeted with some skepticism concerning its practicality. In response, the American Physical Society sponsored a study to assess directed energy weapon technology. One conclusion of the study was that the power needed by lasers to shoot down missiles would have to be *at least two orders of magnitude greater* than the power of existing lasers. Understanding what an order of magnitude is, you see clearly the implications of that conclusion. Note that the study group did not give a precise number for the power needed; nor could they have done so without performing experiments costing millions of dollars and taking years to complete. But using the information at their disposal and their basic knowledge of physics, they were able to calculate the order of magnitude of the required power. This order of magnitude was sufficiently informative, given the number it turned out to be. The emphasis on directed energy beams was dropped soon after the study group's report was published (although the SDI program continued to exist).

Uncertainty

We have so far been discussing poorly known numbers, but some numbers are known very precisely. The charge of an electron, for example, is known to within a fraction of a part per million. Analytical chemists routinely measure masses of chemicals with extremely high precision. But even these numbers are still not exact. We need some way to express numerically how well or poorly a number is known; we need the concept of uncertainty. The uncertainty of a quantity is the range within which we believe this quantity to lie. Conventionally, the uncertainty is usually written after a ± sign following the quantity. For example, your weight might be written as 153 ± 4 lb, meaning that your actual weight is probably between 149 lb and 157 lb. The best value for the weight itself is 153 lb and the uncertainty of the weight is 4 lb. This uncertainty is pretty high, maybe because you are using a cheap old bathroom scale. A better scale (perhaps the kind found in a doctor's office) provides a better value, like 151.8 lb ± 0.1 lb. This value is consistent with the first, but is more precise. Without using any scale at all, you might just guess 160 lb ± 20 lb. Your guess is also consistent with the first and second values, but is now considerably less precise.

The uncertainty of a number is also sometimes called the error of the number, the experimental error, or the error bar. This last term comes from the graphical representation of uncertainty, as shown in Figure 6. Measurement uncertainties can be interpreted within a statistical framework, the mathematical theory of errors. In the natural sciences, the usual custom is to identify the uncertainty with a statistical concept called the standard deviation. Leaving statistical technicalities aside, a rough interpretation of the standard deviation is the following: if a value is measured a large number of times, about two-thirds of these determinations will be within the range given by the uncertainty. In the social sciences, somewhat different customs prevail, but the basic idea is the same (namely, that one attaches a statistically defined reliability to the error range). However, statistical interpretations of uncertainty are only really meaningful when they are based on extremely large amounts of data, which is almost never. So, in practice you are safer to interpret a quoted uncertainty as only a rough estimate unless you have evidence that it's statistically valid.

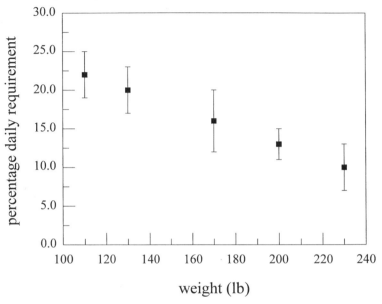

variation of the nutritional content of Wonder Flakes with
a person's weight, using error bars to show uncertainty

Figure 6. Uncertainties are illustrated graphically as error bars. In this imaginary
example, the error bar tells us what the uncertainty range is for each of the percentage daily requirement numbers given by the graph.

Some Interesting Points

Before leaving this topic, there are three technical points worth making:
the difference between precision and accuracy; the nomenclature for indicating an order of magnitude; and the difference between measurement
error and natural variability. Strictly speaking, precision and accuracy are
not the same thing to a scientist, and the distinction is useful. Precision
tells us the range of a set of measurements, that is, how well they agree
with each other. Accuracy, on the other hand, tells us how well these
measurements agree with the correct value. If you weigh yourself on a
doctor's scale, for example, you will get a precise measurement of your
weight, but if you are wearing heavy boots, then you won't get an accurate
measurement (scientists often use the term "systematic error" to refer to
this kind of situation). High precision is no guarantee of correctness.

The method of using the ± sign and a range doesn't work very well
when a number is known only to the nearest order of magnitude. We can

always just use the words "order of magnitude" to indicate this situation, as I've done so far, but that's rather inconvenient. Instead, we have a special symbol, ~, which means the same thing. For example, we can write the nutritional value of Wonder Flakes as ~10% or the typical yearly income of Americans as ~10^4 dollars. I will use this notation in the next section discussing estimation.

The uncertainty of the number representing some quantity can be due to two distinct sources. On the one hand, there may be an uncertainty due to the measurement process even if the number is very well defined (the speed of light, for example). On the other hand, uncertainties may also be due to the natural variability of the quantity itself. If I ask how much vitamin C is in an orange, the answer will involve some uncertainty stemming from the differences among oranges (no matter how precisely I measure the vitamin C content of a particular orange). In a case like this, we want to know both the average (our answer) and also the range (our uncertainty). I find it useful to keep in mind that numbers have uncertainties, and that uncertainty must be considered as we evaluate claims that are made in politics, business, and the media. Always striving to make your own uncertainty estimate for any number you use is a good habit to cultivate.

§3. ESTIMATION

We rarely have the luxury of knowing precise values for numbers needed in the everyday course of life. Two simple examples will show what I mean by this statement. Example 1: Suppose you are traveling somewhere by car and you have a roadmap. You can say *about* where you are on the map and *about* how far you are from your destination. But what you say certainly won't be extremely close (within a mile, say) to the actual distance. Example 2: Now suppose you are reading a newspaper and come across a startling claim, for example, "all of the acidic lakes in the northeastern U.S. can be limed for $500,000 a year" (this is an actual quotation from a newspaper; more on this later). You would need some independent and reliable source of information to determine whether this claim is valid or not. However, in the same way that you can estimate the unknown mileage to your destination, you can estimate the numbers you need to decide whether this claim makes sense. Since we rarely have independent and reliable sources of information at our fingertips, developing skill in estimation can be exceedingly valuable.

Scientists find it invaluable to make rough estimates during the course of their work. Estimates serve as a check on the results of difficult and sophisticated calculations; if the detailed calculation gives a result that seems unrealistic compared to the crude estimate, we have a clue that something may have gone wrong in the maze of details. Making an initial estimate might also tell us whether some idea or plan is feasible and worth the effort of working out the details (or not). The so-called back-of-an-envelope calculation is a standard part of the cultural lore in physics. One person who was famous for his ability to use estimation techniques was the physicist Enrico Fermi. With his quick mind and extensive command of facts, Fermi was legendary for being able to work out complex problems in a simple fashion, and he made it a point to train his students in the same estimation methods of which he was a master. A particularly striking example occurred when Fermi observed the explosion of the first atomic bomb. By dropping bits of paper to the ground and measuring how far they were blown by the wind from the blast, Fermi was immediately able to estimate how powerful the bomb was (in equivalent tons of TNT). His estimate was later verified by careful analysis of instrumental readings.

We have already developed one of the key ingredients we need for skill in estimation: the idea of an order of magnitude. To use this idea effectively, you must be able to make reasonable estimates of the quantities you need. The last key ingredient is the ability to combine your estimates judiciously after you have made them in order to obtain the result you want. In making useful estimates, you also sometimes need to draw on a bit of basic knowledge (i.e., some commonly known facts). An example of what I mean by basic knowledge is the population of the United States, of your hometown, and of the world (rough estimates). A few other examples of useful things to know (roughly) are the density of water, the size of the United States, and how much electrical energy your household uses each month. For any particular fact, of course, you either know it or you don't. Obviously, we can't always happen to know just the information we need, but we can be on the lookout for potentially useful facts and commit them to memory. It's not hard to remember an order of magnitude, which is all you usually need. Let's look at two examples of how to make an order-of-magnitude estimate. The first example demonstrates some of the basic ideas in the technique, while the second example shows how to apply estimation methods in a public policy dispute.

Example 1: Gold bricks

What is the worth of a chunk of gold the size of a typical brick (the kind brick houses are made of)? Just guessing off the top of my head, I have no idea. But with some thought, we can make a pretty reasonable estimate. I know from listening to financial news reports that the cost of gold is roughly $400/oz, that is, $400 for about 30 grams of gold. If we know how many grams are in the brick, we can figure out how many lumps of 30 grams each are in the brick (divide the number of grams in the brick by 30). Each lump costs $400, so $400 times the number of lumps gives us the cost of the brick. Now we need to know how many grams of gold are in the brick (i.e., we need the brick's mass). Your first thought might be to try a direct estimation of this number. If you try to do this, I suspect that your estimate will not be close to the correct number. I'm sure my estimate would not be close. Why is the mass of a gold brick so hard to estimate? Because to make a decent estimate, you would have to lift one in your hand and feel how heavy it is. I don't know about you, but I have never had a brick of gold in my hand. There is a better way to estimate the mass of a gold brick. The only other information we'll need turns out to be the density of gold, which we can either estimate or look up. To estimate, use the density of water (1 g/cm^3) and the fact that heavy metals are an order of magnitude more dense than water (e.g., see chapter 4 and justifiably assume that the densities of gold and mercury have the same order of magnitude). Since the information we want in this case is readily available, though, I just looked it up. I found that the density of gold is about 20 g/cm^3. This means that each volume of 1 cm^3 contains 20 g of gold. The volume of the gold brick is something we *can* estimate (see chapter 16) because it is the same as the volume of an ordinary brick, which we've all seen. The volume of the brick (the number of cubic centimeters it contains) multiplied by 20 then gives us the number of grams in the brick, which we want. As you may have noticed, we've been using the relationship *mass equals density times volume* (see chapter 19). This relationship can be written as an equation, $m = \rho V$ where m is the mass, ρ is the density, and V is the volume.

To proceed further, we need to estimate the volume. We could again attempt to do this directly just by looking at a brick and guessing its volume. Estimating volumes is not so easy, however, and we are likely to be way off in our estimate. An alternative method is to estimate the length of each side of the brick. I find it much easier to estimate a length than to estimate a volume. All we need to do then is multiply the three lengths together to get the volume of the brick. For my estimates of the sides of a brick, I chose 5 cm, 8 cm, and 20 cm. The volume I estimate based on these lengths is then about 800 cm^3. This estimate is probably not very precise, but I'm almost certain that I've gotten the order of magnitude right. We now have all of our ingredients assembled, and we can put them together to find our estimate of the gold brick's worth. Multiply the volume times the density

$$(800 \text{ cm}^3) \times (20 \text{ g/cm}^3) = 16{,}000 \text{ g}$$

to get the mass of the brick. Then divide the mass by 30 to get the number of $400 lumps, which is about 530. Finally, multiply this number times $400 to get $212,000. An equivalent (in fact identical) procedure is to divide $400 by 30 g to get the price per gram, about $13/g, and then multiply this by the brick's mass

$$(16{,}000 \text{ g}) \times (\$13/\text{g}) = \$208{,}000$$

The numbers are slightly different only because I have rounded off differently in the two cases; the difference is not meaningful given the crude estimates we've been using. In fact, the proper procedure is actually to round off the results to $200,000. So we have our answer: the gold brick is worth about $200,000.

There are two especially interesting points concerning this result. One is how large the result is. This is a lot of money, and I doubt if many people would guess offhand that a piece of gold the size of a brick is that costly. That we can get such an unexpected result from numbers we know or can estimate is a testimonial to the power of these methods. The second interesting point is closely related to the first. Even though our number is not very precise, it tells us something important. Why? Because the remarkable characteristic of this result is not the exact amount but rather its order of magnitude. The important information we have learned is also the dependable information (namely, that a gold brick is incredibly expensive even compared to our expectations).

Example 2: Acidic lakes

Let's take a closer look at the example quoted near the beginning of the section ("all of the acidic lakes in the northeastern U.S. can be limed for $500,000 a year"). This quote is from an opinion piece in a 1992 issue of *U.S.A. Today*. The author's purpose is to convince the reader that environmental problems are exaggerated. The quote is taken from a series of claims that are intended to show that acid rain is not really a problem. Liming means adding an alkaline substance to neutralize the excess acidity of the lake (much like the heartburn medicines advertised on TV). We don't have any way to know whether the quote is true or is a lie, but we do have a way to proceed. The question is this: Does this quote make sense?

To progress very far in our thinking, we need to know how many acidic lakes there are in the northeast. Estimating that number would be our first task, and I admit that it would be a tough one. But we are in luck because, in the very same list of claims, the author writes: "There are 240 critically acidic lakes out of more than 7000 northeastern lakes." The author neglects to tell us what "critically acidic" means, but we can certainly assume that 240 must be the smallest number of acidic lakes conceivable. (You may object that if we're worried about the author's lying in the first place, why should we trust this number? If he was lying, though, he'd give us a number that is too small and the real number of acidic lakes would be bigger, so once again we may assume that 240 is the minimum possible number of acidic northeastern lakes.) So we can divide $500,000 (the total claimed cost of fixing the problem) by 240 to obtain the maximum average cost per lake. This should be a good estimate of the cost to neutralize the acidity of a single lake, based on the author's claims. In practice, we can round off to 250 to make the arithmetic easier (obviously we lose no real precision by doing this). The arithmetic is trivial, and we discover that the estimated cost for each lake is about $2000. What have we gained by doing this? We now have a number that we can compare to other numbers we are familiar with (half a million dollars is a little abstract to me). How much of a job does $2000 buy? Compare that amount to some contracting or landscaping job you are familiar with. I paid about that much money to get a new set of drain gutters on my house. Is it reasonable that this amount will pay for a crew of professionals to travel to a lake and perform chemical procedures to neutralize its excess acidity? And remember that this is the *highest* amount it could be.

An Afterthought

Obviously, getting better (more reliable and/or more precise) information is always desirable whenever this is possible. But even a rough estimate can be informative and valuable, and we often learn what we need to know just from the order of magnitude of an estimated number. One of the difficulties of estimation techniques is that there are no simple rules to follow. Not knowing whether we've gotten the right answer and not having necessary information to work with are both somewhat disturbing. The benefits of making estimates, however, are well worth the effort involved both in science and in practical affairs.

§4. THE INTERPRETATION OF NUMBERS

Quantitative reasoning can be a valuable tool in thinking, but if this tool is not used properly, then we might come to incorrect conclusions even when the numbers are right. Let's look more closely at how to interpret numerical information.

Meaning and Context

Numbers have meaning (or at least numbers *represent* something with meaning). The number for the average income of Americans has a meaning in terms of the quality of life of American citizens. The number of people in the world who die of malnutrition has a meaning that is all too obvious. The number of tons of coal burned last year in the United States to produce electricity has a meaning too, but that meaning is less obvious. The convenience of using the electricity; the environmental pollution of burning the coal; the jobs created mining the coal; the importance of the technologies that run on electricity; the profits of the mining and utility companies; and many other factors all contribute to the meaning of that number. We need to have a genuine understanding of a number's meaning in order to use it productively in our reasoning. We should also understand that reducing a concept to a number leaves out some of its meaning. A person's quality of life depends on many other things (relationships, job satisfaction, etc.), not just personal income. The number of people dying of malnutrition tells us nothing about whether the cause is natural disasters or political intrigue. In other words, the use of quantitative reasoning is productive when it's used as part of a more general approach. The meaning of the numbers must be understood and placed into a context that goes beyond quantitative information.

Numerical Comparisons

The information we get from a number can sometimes be made much more meaningful by looking at it in comparison with other numbers. For example, the number of tons of coal burned last year may not tell you much by itself. Compare that number to the number of tons of coal burned 25 years ago; compare it to the amount of known coal reserves in the earth; compare it to the amount of oil or natural gas burned to produce electricity; make these kinds of comparisons and you start to arrive at a deeper understanding of what the number of tons of coal burned last year means. In fact, the ability to make numerical comparisons of this type is one of the greatest advantages of quantification. Comparing things that have not been quantified in this way can sometimes be very difficult. For example, suppose we are trying to decide how to allocate scarce public health resources. Two diseases are both terrible killers of the people who get them; how should we decide which disease to spend more money fighting? On what basis can we compare them? Now suppose that we quantify the number of people who die from each disease in a year, and we discover that one of the diseases kills over a hundred times as many people as the other disease. This quantitative comparison may induce us to make a choice, spending more money on the disease that kills more people. Of course, there are ethical issues and other aspects of the decision that can't be quantified. And once again, we want to keep in mind that we may have lost important information in the process (e.g., how easy or difficult is it to cure each disease?). Despite these considerations, however, the quantitative aspects of a problem like this one are invaluable additions to our thinking process. This ability to make numerical comparisons is one of the unique advantages of quantification.

Relative Quantities

In thinking about the significance of a number, another important consideration is the relative value of that number. By relative value I mean not the number itself, but rather the number compared to some standard or total. As a simple example, let's say that some animal is eight years old. This doesn't tell us very much; in particular, it doesn't tell us anything about whether the animal is old or young unless we know its expected lifespan. If the typical lifespan of this species is forty years, the animal has lived only 20% of its life; whereas, if the typical lifespan is ten years, the animal has lived 80% of its life. Let's take forty years to be the typical lifespan. Then eight years is the actual age of the animal while 20% is the relative age of the animal. As you can see, both of these numbers (eight years and 20%) convey information, each in a different way. As another example, suppose we are trying to decide how important it is to recycle a certain industrially important metal. We find that 25,000 tons of this metal are mined per year. This tells us something of moderate interest, but not really what we need to know. We now discover that this amount (25,000 tons) is 5% of the estimated reserves of the metal in the earth. This number implies that, at the rate we are presently mining it, all the metal will be gone in twenty years. The relative value (5%) is clearly the important thing to know in deciding about this issue.

Precision

Another point to consider in the interpretation of numbers is their precision. For example, suppose one quantity is 3% greater than another, but both are known only to within ±40%. The fact that one is bigger than the other is virtually meaningless in this case. If someone tried to use this 3% difference to make a point in an argument, you would realize that it meant nothing if you knew the precision of the numbers. And if you didn't know the precision of the numbers, you would know that you need to ask for it. Sometimes we know only the order of magnitude of a number, but this knowledge can still be useful. In the example we just discussed, where the precisions are only ±40%, we would still have meaningful information if one number was an order of magnitude greater than the other (as opposed to 3%). The order of magnitude of a quantity can often be estimated, which makes it particularly useful.

Making Numbers Meaningful

The numerical value of a quantity is sometimes not the most important point to note when trying to interpret numbers. Instead, the change in the numerical quantity is sometimes the important point. An example is inflation in the economy, where the numerical value of the dollar is not as important as whether that value is increasing or decreasing, and by how much. The rate of change is also itself a numerical quantity, but further discussion of this point would take us too far afield (see chapters 19 and 20 for more on rates). An additional advantage of quantification is that it allows us to combine and manipulate concepts in a convenient way. We've seen several examples of this already (the solar panel example and the estimation examples). Combining concepts in this way is almost impossible without quantification, and the process allows us to arrive at new and deeper interpretations of the initial concepts. Finally, numbers are sometimes not meaningful because they are too large, too small, or too abstract. In these cases, we can often manipulate the number in some way that makes it more meaningful by bringing it within our experience. As a simple example, recall the number of tons of coal burned per year (discussed above). The order of magnitude for this number is 10^9 tons, a number so large that we have trouble even imagining the amount, and for that reason it's difficult to interpret. If we divide 10^9 by the number of people in the United States, we have the number of tons of coal burned in a year for each person. Since there are about 300 million people (3×10^8) in the U.S., we find that about 3 tons of coal are burned for every person in the country. This result is a more easily imagined and more meaningful number; we get a real sense of how much coal is being burned. Recall that we employed a similar tactic in the acidic lake example.

FOR FURTHER READING

Theory of Errors and Least Squares, by LeRoy Weld, Macmillan, 1916.
Quantification, edited by Harry Woolf, Bobbs-Merrill, 1961.
The Fermi Solution, by Hans Christian von Baeyer, Random House, 1993.

EPIGRAPH REFERENCES: Lord Kelvin, quoted in *From Alchemy to Quarks*, by Sheldon Glashow, Brooks/Cole, 1994, p. 274. R. W. Gerard, in *Quantification*, ed. by H. Woolf, Bobbs-Merrill, 1961, p. 206.

PART III

LARGER QUESTIONS: THE CONTEXT OF SCIENCE

Chapter 9

ULTIMATE QUESTIONS: SCIENCE AND RELIGION

> Science and religion deal with different aspects of
> existence. If one dares to overschematize for the sake
> of clarity, one may say that these are the aspect of
> fact and the aspect of meaning.
> *(Theodosius Dobzhansky)*

THE RELATIONSHIP of science to religion is a broad topic, which has been treated extensively by many profound thinkers. Before starting our brief account, I wish to clarify what is meant by religion here, since it can mean quite different things to different people. I am using the term very broadly to include: organized religions based on well-defined creeds, traditional religious beliefs and experiences in a variety of cultures, spontaneous religious experiences that are not within a particular tradition, and so on. Many kinds of religious experience, based on faith, mystical insight, scripture, and authority, are all included. The traditions and sacred writings of many different cultures are all taken to be equally valid for our purposes here. There is a common element in all these conceptions of religion that is not present in science, and it is this common element and its relationship to science that we will now explore. A number of questions present themselves for consideration: Are there any possible conflicts between science and religion? Are there indeed some conflicts that are inevitable? Do science and religion have any common ground, any synthetic worldview that they share? Or are science and religion completely incompatible, with nothing to say to each other at all? Should religion be considered merely an artifact of human thought and an object of scientific study? Or is it science that should be considered an artifact of human thought, unworthy compared to the eternal verities of religion? These questions don't have simple yes-or-no answers, but I believe that we can say something intelligent about the issues they raise.

§1. SOME FUNDAMENTAL DIFFERENCES

Let's begin by stating clearly the differences in purpose and in method between science and religion. Science attempts to bring coherence to our experiences, whereas religion attempts to infuse our experiences with meaning. This difference is apparent in an example that is at the heart of science and also central to some forms of mysticism: the contemplation of nature. Looking at the wooded bank of a creek, a scientist might perceive the photochemical processes of food production or the flow of energy and material through the ecosystem. A nature mystic looking at the same scene might perceive its deep harmony and sacred beauty. The same person might perceive both of these aspects; there is no contradiction between the two. But they are very different things. No scientific result can give my life or the universe significance; this is not a goal of science. Giving life significance is, on the contrary, one of the very important goals of religion. By the same token, no scientific result can rob our experiences of meaning because science can say nothing one way or the other on this issue. Questions of meaning, which may be central to religion, are outside the scope of scientific discourse.

Scientific statements are ultimately statements about sensory information. This comment is true in spite of the many abstract concepts and terms, not directly observable, to be found in the sciences. Religious statements, on the other hand, are ultimately statements about that which we cannot perceive with our senses. Transcendent beings are not located in space, time, or matter. For this reason, they are not subjects of scientific discourse. The wooded bank of the creek illustrates this point as well. The coherent picture of this scene as an ecosystem is based on observations and measurements. A scientific concept (ecosystem) is exactly what ties the observations and measurements into a coherent picture. The sacred quality of the scene, on the other hand, is not inherent in these observations and measurements. Instead, it's inherent in the experience of the observer.

The discourse of science is always a public discourse. Science is conveyed in terms which can (at least in principle) be precisely defined. Returning to the woodland scene, each niche in the ecosystem and each chemical reaction in the photosynthesis process can be precisely defined and described to someone else. In contrast, religious experiences can be private and are often ineffable, that is, incapable of being expressed in well-defined terms. (An ineffable scientific result is by definition impossible.) The sense of the woodland scene as a sacred place cannot always be communicated to someone else. We can try to put it into words, but the words might be incapable of conveying the real experience.

Finally, the results of science are always tentative results, subject to revision in the light of new evidence or better ideas. Religious statements, in contrast, are not generally intended to be tentative statements subject to revision. Religion is concerned with knowing eternal truths; science is concerned with discovering new and improved ways to understand the world.

§2. Historical Conflicts

For some people, the basic relationship between science and religion is antipathy and conflict. Examples of this conflict are not hard to find. Copernicus' great work was banned, Galileo was tried by the Inquisition, Giordano Bruno was burned at the stake, the use of lightning rods was denounced as impious, Darwin's work was ridiculed and attacked by Anglican clergy, and even today Fundamentalists attempt to persuade schools in the United States to rid the science curriculum of evolutionary theory. A. D. White spent about 800 pages documenting in detail this dismal record of religious interference with science (and that was only through 1895). But was all this really necessary? I see no inherent conflict between science and religion because the two have different purposes and methods (as outlined in §1). If the discourse of religion is concerned with transcendence and meaning while the discourse of science is concerned with understanding the observed behavior of nature, how can there possibly be a conflict? Well, considering the historical record, maybe the question we should ask instead is this: How did it happen that so many conflicts occurred? Presumably, these conflicts occur when religion invades the realm of discourse proper to science, or science invades the realm of discourse proper to religion, or else there is some genuine overlap between the two. Many of the most obvious cases involve religious authorities making statements about nature that (in retrospect anyway) have no particular spiritual content and contradict the evidence of the senses. All of the cases that White details (occurring in Christian Western Europe) fall into this category. Less obvious and less talked about are the cases where scientists overstep the limits of their fields and make statements that have no real warrant in scientific evidence about the spiritual dimensions of humanity. Examples of such statements are these: "The more the universe seems comprehensible, the more it seems pointless" (made by a physicist); "Man must . . . discover his total solitude, his fundamental isolation. He must realize that . . . he lives on . . . a world that is deaf to his music, and as indifferent to his hopes as it is to his sufferings or his crimes" (made by a biochemist). Such statements, which are clearly religious in character, are erroneously passed off as scientific deductions (Mary Midgley has written particularly lucid work on this point).

The question remains: Is there some area of legitimate overlap between science and religion wherein conflict (or cooperation) may occur? We'll take up this question at the end of the chapter. Meanwhile, let's look at two final points concerning science/religion conflicts. The first point is this: Conflicts don't involve abstractions such as "science" and "religion." Conflicts involve people, scientists and religious authorities. As such, there are questions of political influence and power as well as the involvement of other participants in the society (e.g., scholastic academics). There are many complications in the real cases that have occurred, the details of which are best studied by historians and sociologists. It's worth keeping in mind, though, that philosophical conflicts occur in a social context. The second point is that the Christian West had particularly bad conflicts compared to other cultures. I think this happened for two main reasons. One reason is that the churches adopted superfluous views about nature (such as geocentrism) and incorporated these into the fabric of their orthodoxies. In contrast, for a teaching such as Zen Buddhism, most scientific questions are simply irrelevant, and so conflict in this area is not possible. The other reason is that there were already secular elements in European culture at that time to which science and the scientists could belong. In contrast, science in the Islamic world of the twelfth century was integrated into a cultural milieu grounded in Islam, which permeated all aspects of life; in other words, science could not conflict with religion because science did not exist apart from religion.

§3. Faith, Experience, and Meaning

Religion and science do have some characteristics in common, and looking at these commonalities gives us a deeper understanding of their differences. Many simplistic accounts state that religion is based on faith while science is based on experience. As we'll see, however, both faith and experience are intrinsic to both religion and science, but in very different ways. Meaning is also central to both religion and science, but once again the sense of the word "meaning" is quite different in the two cases. The role of faith in religions is well known, but in what sense does faith play a role in the sciences? Isn't science based on skepticism, the opposite of faith? Concerning any individual results, claims, or theories, the answer to this is yes. But in order to do science at all, scientists must have an underlying faith that nature is subject to scientific understanding, that nature is in some sense lawful, rational, and orderly (I only mean by this that nature is not totally capricious; this is not a strong claim in any way). As the chemist/philosopher Michael Polanyi has emphasized, scientists must also have an implicit faith that the overall premises and methodologies of science are valid. Religious faith is of a different sort, but religious faith itself comes in several varieties: faith in the truth of particular dogmas or scriptural writings; faith in the pronouncements of religious leaders or authorities; or faith in the reality of a lived religious experience. Finally, one must at least have faith in the integrity of one's overall worldview, which might be grounded in religion. In this last case, we are getting close to the kind of faith discussed above in connection with science.

What role does experience play in science and in religion? In science, experience means experience through sensory perception (and its extension by instrumentation). This is essentially what we mean by empirical observation. Religious experiences, on the other hand, are not primarily sensory experiences (the senses may have some involvement, but ultimately the experience is not one of sensory perception). Instead, they are internal experiences. The revelatory gnosis of the Christian mystic; the Enlightenment of the Zen master; the initiation at Eleusis; the moment when the Sufi adept finally understands the hidden meaning of things; all these are solitary and internal experiences that do not depend on the evidence of the senses and cannot be communicated by words in the usual fashion. For both of these reasons, such experiences play no role in the discourse of science. Now, can we form some judgment as to whether one or the other of these very different and incommensurable kinds of experience is more valid or real than the other? This is obviously a matter of great controversy, which I will refrain from entering. Instead, let me quote at some length from two sources with opposite points of view. From E. O. Wilson: "The enduring paradox of religion is that so much of its substance is demonstrably false, yet it remains a driving force in all societies. . . . The individual is prepared by the sacred rituals for supreme effort and self-sacrifice. Overwhelmed by shibboleths, special costumes, and the sacred dancing and music so accurately keyed to his emotive centers, he has a 'religious experience.' " And from A. S. Eddington: "Are we, in pursuing the mystical outlook, facing the hard facts of experience? Surely we are. I think that those who would wish to take cognisance of nothing but the measurements of the scientific world made by our sense-organs are shirking one of the most immediate facts of experience, namely that consciousness is not wholly, nor even primarily a device for receiving sense-impressions." There is no right answer on this issue. We all must arrive at our own conclusions.

We turn now to the role of meaning in science and in religion. I previously wrote in §1 that meaning is the province of religion and has no role in science, but I was using the word "meaning" in a particular way. Bringing order and coherence to a set of otherwise unrelated facts and observations certainly gives these facts and observations meaning. This is one of the important things that science does, and in this limited sense meaning is crucial to scientific thinking. However, meaning in the sense of ultimate significance or metaphysical purpose is alien to scientific thinking; whereas this kind of meaning is relevant to (perhaps even crucial to) religious thought. Consider as an example an epidemic in which many thousands of people die. If we can identify a microorganism causing the disease and determine how it is spread, we have provided a scientific meaning for the seemingly random patterns of death. But to ascertain the spiritual meaning of these deaths, if there is any, requires a different mode of thinking. Confusing these two different roles of meaning can cause serious problems in the relationship between science and religion. To attach great metaphysical significance to the "fact" that the earth is the center of the universe by concluding that humanity is thus central to the divine plan is a grave mistake. Why? Because the position of the earth is an empirical question, while the position of humanity in the divine plan is clearly a religious question. When the two were joined together in medieval Europe, misunderstandings were caused that apparently still exist. It turns out, of course, that the empirical answer to the empirical question is that the earth is not the center of the universe. We now have people making the naive assertion that because the earth is not central to the universe, we can conclude that humanity is thus peripheral in some metaphysical sense. This confusion has been widespread (Newtonian physics was mechanistic, so God became a clockmaker). If there are any useful connections between levels of meaning in science and in religion, they must be made carefully, thoughtfully, and with circumspection.

§4. MUTUAL INTERESTS?

Now that we understand the important differences between science and religion, we are in a position to ask these questions: Do they share any common ground at all? Do science and religion have anything of interest or value to say to each other? Or are they in fact mutually exclusive categories with no possibility of overlap? At least one tradition asserts that science and religion have a meaningful overlap in the following sense: If nature is the embodiment of some spiritual divine presence, then the study of nature (i.e., science) is a way to better know this divine presence. This seems to have been the position of Galileo, for example, who writes in his letter to the Grand Duchess Christina that "God reveals Himself in no smaller measure in the phenomena of nature than in the sacred words of Scripture." The study of nature as a means of spiritual enlightenment is a tradition going at least as far back as Pythagoras; it was prominent in the thinking of Roger Bacon, Isaac Newton, and Albert Einstein; and it's still held by a number of scientists alive today.

A counterargument to this tradition might be that there are scientists who are atheists, but I think that this misses the point. After all, a Christian scientist and a Hindu scientist might both study science in the same way (as would an atheist scientist), and they both might feel that their study of science deepened their spiritual comprehension of the world (they'd part company with the atheist here). But each one could arrive at a different spiritual understanding of the science they shared. This difference is entirely legitimate, based on the premises we've discussed. Science cannot dictate religious conviction in any public and/or necessary way, but science can certainly inform the religious convictions of individuals on a personal level. The atheist, of course, is also free to reject on a personal level any such religiously informed understanding of science. But nobody can claim that this religiously informed understanding of science is wrong in any fundamental way, as long as the science itself is valid. The fact that science does not require any nonmaterial entities doesn't logically exclude their existence. It is impossible to prove that God exists based on scientific grounds. (I have ignored the voluminous and pointless literature on this question; I hope what I did write makes clear why the statement is correct.) For identical reasons, it is equally impossible to prove that God doesn't exist based on scientific grounds. There is a famous story that when Napoleon asked Laplace why he left God out of his book *Celestial Mechanics*, Laplace replied "I had no need of that hypothesis." Laplace was correct, of course, but that doesn't tell us anything about God. It only tells us something about celestial mechanics. In other words, science can get along quite well without religion; but this fact does not imply that science has somehow invalidated religion, nor does it imply that science and religion are incompatible.

An interesting modern development is the attempt to interpret the religious significance of quantum mechanics. A number of books have been written on this theme, perhaps the most prominent being the one by F. Capra. The basic idea is that quantum mechanics can be interpreted as indicating that all parts of nature are fundamentally interconnected, an idea that is similar to the teachings of several Eastern religions. While this comparison is certainly of some interest, we should keep in mind that the meaning of interconnectedness may not be identical in the scientific and religious contexts. Moreover, it is somewhat dangerous (as I've already discussed) to tie one's religious truths to scientific theories, because these theories are liable to change as our knowledge and understanding progress. The teachings of Eastern religions, in contrast, have been stable for thousands of years. Other attempts to amalgamate science and religion need to be considered in the same manner. Perhaps the most well known of these is the work of Teilhard de Chardin, blending together ideas of evolution and Christian theology. Such systems of thought can succeed in providing personal meaning to individuals, but extreme care is called for to avoid violating the legitimate premises of either science on the one hand or religion on the other.

Turning to a different topic, values and ethics certainly constitute an area where religion and science intersect. This topic, however, is so important that I have treated it separately in a chapter of its own (chapter 11). A number of specific proposals for dialogue between science and religion have been made in the literature, but I won't deal with them here in any detail. Instead, I will suggest one last broad area where science and religion can meet: the question of what it means to be human. This question is asked both by science and by religion, each in its own way. We have, unavoidably, an area of mutual interest. It remains to be seen whether it's an area of incompatibility. I suggest that science and religion need not be incompatible on this issue if neither transgresses its appropriate province. Whether science and religion might even enrich each other on the issue of what it means to be human also remains an open question.

FOR FURTHER READING

The Warfare of Science with Theology, by A. D. White, George Braziller, 1955 (originally published 1895).

The Varieties of Religious Experience, by William James, Longmans, Green, 1902.

Science and the Unseen World, by A. S. Eddington, Macmillan, 1929.

Science, Faith, and Society, by Michael Polanyi, University of Chicago Press, 1964 (originally published 1946).

The Challenge to Religion, in *The Challenge of Science,* by George Boas, University of Washington Press, 1965.

Science and Faith, by Siegfried Muller-Markus, in *Integrative Principles of Modern Thought,* edited by H. Margenau, Gordon and Breach, 1972.

Criteria of Truth in Science and Theology, in *Revolutions and Reconstructions in the Philosophy of Science,* by Mary Hesse, Indiana University Press, 1980.

The Tao of Physics, by Fritjof Capra, Shambhala, 1983.

Evolution As a Religion, by Mary Midgley, Methuen, 1985.

EPIGRAPH REFERENCE: Theodosius Dobzhansky, quoted in *Evolution as a Religion,* by Mary Midgley, Methuen, 1985, p. 13.

Chapter 10

MORE PRACTICAL QUESTIONS: SCIENCE AND SOCIETY

> Science, by itself, provides no panacea for individual, social, and economic ills. It can be effective in the national welfare only as a member of a team. . . . But without scientific progress no amount of achievement in other directions can insure our health, prosperity, and security.
>
> *(Vannevar Bush)*

CENTURIES AGO, Francis Bacon eloquently expressed the idea that science would contribute many practical benefits to society in general. Bacon hoped that society would in turn devote resources to science so as to hasten scientific progress. I believe it's fair to say that Bacon's vision has come to pass, and this chapter is a brief look at the current situation. Science has brought problems as well as benefits to society, but I don't dwell much on the problems here because they are dealt with more fully in chapter 11. This chapter simply presents an overview of some basic topics dealing with how science relates to the broader society.

§1. TECHNOLOGY AND SCIENCE

The relationship of science to technology is unclear to many people. Even scholars who study this relationship don't always agree with each other. There are a number of points to discuss, however, on which a broad consensus has formed. One common misconception is simply to equate science with technology, to assume they are both the same. All of the technological gadgets and wizardry we are so accustomed to seeing (spaceflight, ultrasound imaging, computerized special effects on film, microwave ovens, laser eye surgery, and so on), all these things are taken to be science. But these things are technology, which is not the same as science. Science is a way of understanding the world (including both the methods used to acquire knowledge, and also the facts and theories that make up our current worldview). Technology, on the other hand, is a way of controlling the world, a set of tools that we can use to make things happen as we wish. So science and technology can sometimes be separate and unrelated. For example, many important improvements were made to the steam engine without any real scientific understanding of those improvements. The classification system for plants and animals was devised by Linnaeus without either using or influencing any technology. But I think that these examples are probably the exceptions and not the rule.

More typically, science and technology are highly intertwined. Advances in scientific understanding enable the development of new technologies. An example is the discovery of electromagnetic induction by Michael Faraday, which ultimately led to the development of the huge dynamos that supply our electrical power. New technologies have also led to scientific advances. The first telescopes were discovered by accident and improved by trial and error (i.e., they were not applications of optical knowledge, which did not yet exist); these telescopes played a pivotal role in the rise of modern science (see chapter 5). Over the course of time, the relationship between science and technology has become steadily more important. In modern times, this relationship is vital. The entire microelectronics revolution, which has so powerfully affected our lives, is based on a fundamental understanding of the physics of semiconductor crystals (see chapter 1). Lasers are now central to such widely varying applications as compact disc players and retina surgery; the invention of the laser evolved directly from studies of atomic and molecular energy levels. But both of these examples also show that the relationship between science and technology is a two-way street. Computers, which were first enabled by the advance of science, are now indispensable in the laboratory and are used to make calculations that were previously impossible. Lasers have opened new frontiers of precision and sensitivity in the study of atomic and molecular energy levels (among many other areas), the study of which had once led to the beginning of the laser itself. Science and technology depend on each other for their continued vitality.

Technology is sometimes referred to as applied science, but this is not quite right. Although modern technology greatly depends on the results of modern science, there is also a strong element of craft knowledge and lore (tricks of the trade) that is needed. This blending of scientific understanding with technical know-how is what most characterizes technology in the modern era (although various factions sometimes stress one aspect or the other). Applied science is itself a term that is used in several different ways. The term might be used to describe the employment of scientific results to achieve some end (example: using total internal reflection to transmit light through a thin glass fiber). Alternatively, the term might be used to describe the scientific study of some process or phenomenon of practical importance (example: studying the optical properties of glasses in order to make more effective glass fibers). Finally, this same term can also be used to describe the development of innovative engineering techniques based on current science (example: creating novel glass processing methods to manufacture the improved fibers). All these different shades of meaning for applied science are legitimate and contribute to technological progress (to fiber optics communication technology, in this example). It's often important, however, to bear in mind which shade of meaning is being used in any particular discussion.

We are tempted to think of research and development as a simple linear chain of discrete activities, with each link leading to the next one: pure science—applied science—engineering—technology. But all of these activities overlap all of the others in complex ways, so our simple discrete chain is not a very realistic picture. Why does it matter very much how we think about these things? One reason it matters is because science itself affects society only marginally, but science-based technology affects society in extremely powerful ways. It has been credited with major blessings (better health, material prosperity, etc.) and equally important curses (environmental problems, more deadly munitions, etc.). Technology is a key link connecting science and society. Society, through public policy decisions, is responsible for the uses to which technology is put, and also is responsible for the allocation of resources to our research and development enterprise.

§2. PUBLIC FUNDING OF SCIENCE

The prophet of public funding for science was Francis Bacon. In his fictional work *New Atlantis* written in 1624, Bacon envisioned a utopian society that supported systematic scientific research to unlock the secrets of nature and systematic applications of this knowledge to produce practical benefits. Although Bacon's work may have influenced the founding of the Royal Society, the King only gave his good name to the endeavor; he did not give any money. Bacon was far ahead of his time. For the next three centuries, scientific research was done on low budgets. Occasional wealthy patrons provided support, a few wealthy amateur scientists supported themselves, universities contributed a little bit, and a few private foundations funded research on a small scale. Government funding was mostly limited to some practical research in fields like agriculture (with a few exceptions, like the Smithsonian Institution). As technology became more commercially important, companies like Bell Telephone and General Electric founded industrial laboratories to perform basic research in areas of interest to them.

So the situation remained until World War II. The pressing needs of the war required quick results in areas such as the development of radar, antibiotics, and the atom bomb. Given the emergency situation, money was no object. Government officials recognized the valuable contributions of science during the war, and Franklin Roosevelt asked the head of the scientific war effort, Vannevar Bush, to write a report on the role of government in postwar science. This influential report, entitled *Endless Horizons*, recommended large scale funding of scientific research by the government and led to the creation of the National Science Foundation. Since that time, the federal government has dominated the funding of science. I'm not as familiar with the particular historical circumstances in the other industrialized countries of the world, but all of them have devoted significant amounts of money to the support of science in the postwar period. While the motivation of a scientist is often simply curiosity and a desire to understand nature, the motivation of society to fund science (through its government) is generally more practical. Of course many citizens are fascinated by the new results of scientific research, but I doubt that this alone would justify the amount of tax money spent on research and development. Economic prosperity, military security, and better health are more often cited as reasons for societal support of science.

In the decades following World War II, the arms race and the space program rivalry resulting from the Cold War served to stimulate governmental funding of science. The booming economic climate of this period also stimulated the growth of scientific funding (leading to rapid scientific progress). The more stagnant economic times that followed caused a slowdown of this growth and a renewed debate over the justification for science funding. Several large multibillion dollar scientific projects were all started around the same time, unfortunately on the verge of a recession. One of these projects, the superconducting supercollider, was cancelled while still under construction (the first time anything like that had ever happened). The countries of Europe also found it more difficult to keep up funding for science, and the situation was worst of all in the remnants of the former Soviet Union. Finally, the end of the Cold War removed yet another rationale for the support of research.

The success of Japan in fostering the commercial development of technologies had encouraged a new element in the debate over public funding of science in the United States. There seemed (in the United States) to be some impediment to the use of basic scientific research results in commercial applications. The Japanese government took a more active role in promoting such technology transfer, and voices were raised in favor of adopting a similar policy in the United States. Similarly, a shift in emphasis away from basic scientific research in favor of applied science and engineering was advocated. This position was rather controversial, however, finding supporters and detractors in both the scientific and the political communities. Meanwhile, the amount of basic research done in the major industrial laboratories steadily decreased. As I write this, the situation remains unsettled. Funding for pure scientific research has remained at roughly the same level for about a decade or so. Government spending is highly impacted by the stringent fiscal climate imposed by simultaneous tax cutting and budget balancing. Funding for scientific research seems to enjoy support in both the general public and the political establishment, but unresolved issues remain concerning the proper level of such funding and how it should be allocated. One positive outcome of the science funding controversies is that scientists are now making greater efforts to explain the results of their work and its benefits to society.

§3. Science as a Social Institution

About ten million people in the United States have degrees in a science or engineering field, about a half-million with doctoral degrees. In the natural sciences, these numbers are roughly three million total and a quarter-million doctoral degrees. Although we have a long way to go before the gender and ethnic makeup of people in the sciences mirrors that in the general population, we have made some progress in the last few decades. The common stereotype of the scientist as an old (dull) white male is contradicted by the wide variety of people who go into the sciences. Nor do people who become scientists have any single type of personality. Introverts, extroverts, competitive people, cooperative people, nice people, and obnoxious people all sometimes become scientists. Curiosity and flexibility are probably more common traits in scientists than in the general public, but even these traits vary widely from one scientist to another. And just as there is no single personality type, there is also no single monolithic culture in the sciences. Some subfields of science foster a culture that is highly competitive and hierarchical, while others are much less so.

A common element in all forms of science, considered as a social activity, is communication. A discovery or theory doesn't become incorporated into science until it's communicated to the scientific world. The primary means of communication in the sciences are journals and meetings; books tend to play a larger role in the later stages of the process as knowledge becomes more consolidated, while electronic mail is becoming more important for the instantaneous transmission of information about work in progress. Scientific meetings are a time-honored tradition, bringing scientists together to exchange insights and information through formal presentations and informal discussions. Scientific journals are the primary medium for the formal announcement of most results. Journals and meetings are often sponsored by scientific professional associations, another tradition that can trace its roots back to the time of the Accademia dei Lincei (Italy, 1600), Royal Society (England, 1662), and Academie des Sciences (France, 1666). Scientific journals are a very specialized communications medium, designed for experts to present technical results to other experts. No one reads these journals for their entertainment value. Readers of scientific journals do, however, want some assurance concerning the scientific value of what they read; time spent studying sloppy work or half-baked ideas is time wasted. To assure the quality of the work they publish, journals use the charming old custom of peer review (peer review is also used by funding agencies to choose which research proposals are funded). Any article submitted for publication is first sent to experts in the field for critical scrutiny. These reviewers, who remain anonymous, can demand revisions or even bar publication. Only work that is approved by these referees is published in scientific journals. Not surprisingly, peer review is a bit controversial. Very innovative ideas and unexpected results tend to get selectively filtered out, making peer review a force for conservatism in science. Since scientists are humans, there is some risk of cronyism and deference to reputations. Sloppy referees do occasionally overlook mistakes. Despite these flaws, peer review is solidly entrenched in the sciences. The system is far from perfect, but basically it does work, and most scientists consider it far better than the alternatives.

Like any human institution, science is affected by the motivations and foibles of its practitioners. One motivation of scientists, of course, is simply curiosity and a desire to find out more about how nature works. But scientists are also motivated by a desire for rewards, as most people are. A difference between science and many other professions is that money is only secondarily important as a reward; the main desideratum of scientists is more often recognition and fame. A problem can arise when the desire for rewards (or fear of failure) overcomes honesty and a scientist engages in fraud. Despite a number of well-publicized cases, fraud is quite infrequent in the sciences (the social penalty is very high). Another example of human failings creeping into the rationalist paradise of science is the social and political negotiations that occur. For example, a youthful scientist might back an older colleague who has more influence and authority in the social system rather than a junior colleague who has more convincing evidence. Sociologists of science have recently taken an interest in this kind of activity and tried to document it. Their studies have yielded some interesting insights, but the degree to which these social interactions actually affect scientific results has probably been exaggerated (see chapter 15). On the other hand, social factors probably do influence greatly the success or failure of some individual scientists, regardless of the quality of their work.

The community of scientists interacts with the broader society on a variety of levels. We have already discussed technology, and the profound impact that science has on society through its applications. We've also discussed the resources that society provides to the scientific enterprise. In chapter 11, we see that various questions concerning values and ethics link science with the rest of society. Yet another important link between science and society is the educational undertaking; one responsibility of scientists is to ensure that the citizenry has an appropriate understanding of science. The legal system is also an area where science and society intersect; for example, DNA evidence in criminal trials and the use of epidemiology in toxic substance liability suits. Many aspects of the science/society interface (technology, education, law, values) are relevant to public policy controversies that have an important scientific dimension. Examples of such controversies are legion: acid rain, nuclear disarmament, genetic engineering, biodiversity, radioactive waste storage, ozone layer depletion, environmental pollution from pesticides, and so on. In the last section, we'll look more closely at two of these controversies, global warming and the health effects of power lines.

§4. Public Policy Controversies

Global Warming

The basic science underlying the global warming controversy involves a process known as the greenhouse effect. The sun's rays strike the earth and warm it. As the temperature of the earth rises, the earth also radiates away some of its heat (in the form of infrared rays); the higher the temperature, the more infrared radiation. The balance between heating (due to the incoming sunlight) and cooling (due to the outgoing infrared radiation) results in some equilibrium temperature for the earth. If we had no atmosphere, more infrared rays would escape into space and the earth's temperature would be much colder. Instead, certain gases in the atmosphere (such as water vapor and carbon dioxide) block some of the outgoing infrared rays, trapping their heat energy and further warming the earth. These gases operate in much the same way as the glass in a greenhouse, letting in the light (with its energy) but not letting out the infrared; hence the name, greenhouse effect. The controversy concerns the effects of increased carbon dioxide in the atmosphere due to the burning of fossil fuels. Since carbon dioxide is one of the greenhouse gases, increasing its concentration has the effect of trapping more heat. The worry is that this might cause a rise in the average global temperature of the earth; no one knows the effect of such a global temperature rise on the world's climate patterns. If the American midwest became prone to drought or the antarctic ice melted significantly, the results might be disastrous. There is no controversy over the fact that CO_2 levels in the atmosphere are rising; data for this increase is very clear. The controversy is over what effect, if any, these CO_2 levels will have on the global climate. The related political controversy is whether to legislate limits on the amount of CO_2 emissions allowed.

The scientific basis for the greenhouse effect is clear enough, but there are two good reasons for the continuing scientific controversy. One reason is that the atmosphere is a very complicated system, which makes it difficult for us to understand all of the processes well and to make accurate predictions. The other reason is that we have poor data on the mean global temperature and its history, which makes it difficult to judge whether global warming has already occurred or not. In addition, the political, economic, and ideological dimensions of the issue all contribute further to the controversy in important ways. Let's first look at the complexity of the atmospheric processes in order to find out why it's not obvious whether we'll have global warming. If we try to model the effects of increasing CO_2 on temperature, we must include several important feedback processes in our model (see chapters 6 and 21). For example, increasing the air temperature will increase the humidity, and water vapor is also a greenhouse gas. In this case we have a positive feedback loop, which will make the problem worse. However, increasing the water vapor in the air will also tend to increase the cloud cover, and clouds reflect back the incoming sunlight. This reflected sunlight never reaches the ground to warm it. So in this case we have a negative feedback loop, which tends to restrain global warming. A number of these feedback mechanisms are at work (involving snow cover, ocean temperatures, and even the world ecosystem), although the clouds are probably the biggest feedback effect that we don't understand well. In order to make an accurate prediction of the temperature change wrought by an increase in CO_2, our model must include these feedback mechanisms properly.

Climate researchers use extremely sophisticated models of atmospheric dynamics. These general circulation models, as they are called, include the effects of changing latitude and longitude, changes with altitude, convection of heat due to wind circulation patterns, and a variety of feedback mechanisms. The variables calculated in these models include temperature, barometric pressure, and humidity. As you can see, the complexity of climate models is astounding. They need to be "run" for long periods of time on supercomputers, especially if we want projections for decades into the future. This very complexity is what makes it difficult to assess the accuracy of the projections produced by these models. How can we know whether we've left some important piece of the problem out of our model, significantly altering the final result? General circulation models have been employed to find out how including an increased concentration of carbon dioxide in the atmosphere affects the mean global temperature over a time period of about fifty years. Different models, using various approximations and assumptions for the dynamics of the atmosphere, naturally yield somewhat different results for global warming. Climate researchers, under the auspices of the world's governments, have convened meetings to consider all of the available results and arrive at a consensus on the most reliable overall projection. The consensus is that a doubling of the CO_2 level will result in a global temperature rise of several centigrade degrees, but the uncertainty (see chapter 8) of this projection is almost as large as the rise itself. Attempts have been made to check the accuracy of the climate models by using them to calculate the greenhouse gas warming for the last fifty years and compare the calculation to measured temperatures. Unfortunately, this process is not as easy as it sounds. We need the average temperature over the entire world, but very few good measurements exist for the nonindustrialized areas and the oceans; in other words, we have little data for most of the world! Where we do have measurements, local changes (warming due to urbanization, for example) tend to dominate. On the global level, short-term changes (such as worldwide cooling caused by volcanic eruption of particulates blocking the sunlight) further complicate the situation, thwarting our attempts to make meaningful comparisons. In the end, we have no way of knowing whether global warming has already occurred or not, based on the recent temperature history we have available.

Since the historical temperature record doesn't give us an unambiguous answer, and the results of climate models can't tell us with absolute certainty what will happen, what can we conclude? Although this question is a scientific problem, it also has much broader implications. What public policy options should be adopted concerning fossil fuel burning? Economic, political, and ideological considerations enter here. If we try to stay as close to scientific considerations as possible in making our assessment, we need to think in terms of probabilities (see chapter 7). The uncertainties of the climate models prevent us from knowing exactly what the effect of increased greenhouse gases will be; the consensus which emerges from looking at all of the modelling work, however, suggests that nonnegligible warming is reasonably probable. This conclusion is the best that science can do for the present, and society needs to make decisions based on this probability (which is meaningful) rather than on certainties that we can't have.

Power Lines and Public Health

In the late 1970s, reports began to surface linking various health problems to low-frequency electromagnetic fields. The AC electrical power we use all the time (from the wall socket) has a frequency of 60 cycles per second, which is low. In other words, we are exposed almost continuously to this kind of weak low-frequency field. If the fields due to electrical power lines have an adverse effect on our health, then this is a major issue. Since the utility industry (and by extension the rest of our economy) has billions of dollars invested, and since we depend so greatly on our electrical technology, the issue is certain to generate controversy. The basis for the idea that power line fields are dangerous comes from epidemiological studies. Epidemiology tries to trace the cause of a disease by finding correlations between the incidence of the disease and some other factor (which is taken to be the cause). For example, if all the people who got food poisoning during an epidemic had eaten the crab cakes at a certain restaurant, we can probably conclude that these crab cakes were responsible. But epidemiological work is rarely this easy or unambiguous in real life. The correlations are often quite weak, and detailed statistical analysis is needed to find out whether such weak correlations are telling us something real or are just small chance happenings. To use the standard jargon, we want to know whether the correlations are statistically significant.

The early studies, looking for correlations between childhood leukemia and proximity to electrical transformers, were suggestive but not conclusive. These results stimulated further research, and a number of studies were undertaken. Some of the studies showed correlations and some did not, resulting in a confused picture for a while. During this unsettled period in the epidemiological research, the media unfortunately presented a somewhat distorted picture of the situation (including an inflammatory series of articles by P. Brodeur). Meanwhile, the epidemiology work continued, including a number of studies looking at workers with high occupational exposures to low-frequency fields. One major problem with all of these studies became apparent as work progressed. Very few good measures of the "dose" were possible in any of this work, especially compared to a control group reflecting the general population. No one really knew what the field strengths or the durations of exposure were for either the studied groups or the controls. Assumptions made about exposure based on occupation or on proximity to a high-voltage power line often turned out to be questionable.

Another problem was the lack of a plausible model for how these fields might cause cancer or other diseases. As we've seen (chapter 7), correlations alone don't imply cause, and we would like to know what the mechanism is (or could be) in a case like this. The relationship between power line fields and health effects is particularly puzzling because these fields are very weak inside the body. The naturally occurring fields produced by our own physiological functions are much bigger (typical values for the field across a cell membrane, for example, are over a million times larger than the fields induced across such a membrane by external sources like high voltage lines). Even the weak random changes in fields to which our cells are constantly exposed (due to thermal noise, for example) turn out to be larger (by about a thousand times) than the comparable fields due to power lines. Some laboratory studies of the effects of low-frequency electromagnetic fields have been done, mostly using in vitro cell cultures. The dose can be varied easily in these experiments, and comparisons made to a field-free control sample. Some interesting effects have been observed, but usually at very high field strengths. At the low fields comparable to the situation we're discussing, only a few studies have resulted in any observable effects. Those effects tend to be only marginally above threshold and difficult to reproduce. In any event, none of the claimed effects are able to account for any adverse health consequences (such as cancer) due to power lines.

We now have over a hundred epidemiological studies on this issue. Taken collectively, these studies show virtually no adverse health impact of exposure to low-frequency electromagnetic fields. A few individual studies have indicated weak effects, which has kept the issue alive as a public controversy. But these effects are rarely replicated in later studies; apparently, they are just random shifts away from the average. We actually expect that random variations of this sort will happen (like getting five heads and one tail when flipping a coin six times). Our reliable source of information is the entire set of studies taken as a whole, which doesn't demonstrate any problem. What can we conclude from all this? Once again, we need to think in terms of probabilities. We can't have an absolute certainty that the low-frequency fields cause no harm at all to our health. I also can't be absolutely certain that I won't be struck down by a falling meteorite when I leave my house tomorrow. But neither of these propositions is very probable. Given the feeble epidemiological evidence, negative laboratory studies, and lack of any plausible mechanism, the scientific consensus is that power line fields are not dangerous to our health. Not every single scientist agrees with this consensus view, and the public controversy is surely far from settled. But for the vast majority of scientists, the scientific controversy is in fact virtually settled. The proposition that low-frequency fields pose a public health threat is so improbable that we can consider the issue moot.

FOR FURTHER READING

Endless Horizons, by Vannevar Bush, Public Affairs Press, 1946.

Selected Writings of Francis Bacon (includes "New Atlantis"), Random House, 1955.

The New Scientist, edited by P. C. Obler and H. A. Estrin, Doubleday, 1962.

Science, Man, & Society, by Robert B. Fischer, W. B. Saunders, 1975.

"Annals of Radiation" by Paul Brodeur, in *The New Yorker,* June 1989, p. 51.

Science, Technology, and Society, by Robert E. McGinn, Prentice-Hall, 1991.

Confronting Climate Change, edited by Irving M. Mintzer, Cambridge University Press, 1992.

"Magnetic Fields and Cancer" by Edwin L. Carstensen, in *IEEE Engineering in Medicine and Biology,* July 1995, p. 362.

EPIGRAPH REFERENCE: Vannevar Bush, *Endless Horizons,* Public Affairs Press, 1946, p. 42.

Chapter 11

DIFFICULT AND IMPORTANT QUESTIONS:
SCIENCE, VALUES, AND ETHICS

> Our zealous endeavor to create a "value-free" science—
> which seems so essential a requirement of objective scientific
> method—has meant simply that the values dominating our
> thinking have retired to the arena of our underlying presuppo-
> sitions, where they can maintain themselves against critical
> appraisal by being so completely taken for granted that no
> one's questioning attention is focused upon them.
> *(E. A. Burtt)*

> We [scientists] produce the tools. We stop there. It is for you,
> the rest of the world, the politicians, to say how the tools are
> used. The tools may be used for purposes which most of us
> would regard as bad. If so, we are sorry. But as scientists,
> that is no concern of ours. This is the doctrine of the ethical
> neutrality of science. I can't accept it for an instant.
> *(C. P. Snow)*

QUESTIONS concerning the relationship of values and ethics to science are extremely important because science affects humans so powerfully. Science affects people in several ways, both directly and indirectly. Some examples of the influence of science are these: the profound changes in worldview that have accompanied major scientific revolutions; the effect of movements like behaviorism and sociobiology on humanity's self-image; and the indirect effects of science resulting from the technologies enabled by scientific discoveries. Both the direct effects of science and its indirect effects have implications in the realm of values and ethics. The relationship between science and values is not simple, despite the many simple statements that are made. One such statement (often heard) is that science and values are unrelated because science is objective and value-free. In contrast, we sometimes hear that the study of science is evil because science is soulless and mechanistic (and/or it produces destructive and powerful technologies). At the opposite pole, other writers maintain that the study of science is an unmitigated good because science leads to the truth and produces material prosperity. We'll subject all of these simple claims to critical scrutiny in this chapter.

§1. THE INHERENT VALUES OF SCIENCE

The community of scientists shares, as a group, certain values. I am not talking about subtle unexamined biases here, just simple virtues like honesty and curiosity. Now honesty is usually considered a virtue by almost everyone, not just in the sciences. Curiosity, however, is not necessarily considered a virtue by all groups in our culture. Both of these values (as well as the others we'll discuss) are particularly highly regarded in the sciences. Let's consider some of these scientific values and their implications in more detail.

Free Flow of Information

The scientific community is in general opposed to secrecy and isolation. Scientific progress depends on the free and unimpeded flow of information from one scientist to another. If a scientist doesn't publish her results, that is, share these results with the wider community, then the results cannot contribute to a progressively more refined understanding of nature. A commitment to open communication of results is one of the bedrock values of science. Yet, scientists working in the military and in industry often need to keep scientific information secret. Their values as scientists are in conflict with their values as members of another societal institution. Each individual scientist must come to terms with this conflict, and the larger community must resolve issues as they arise. A famous example of this conflict occurred during the Manhattan Project (the effort to develop an atomic bomb), when General Groves (the military project director) tried to compartmentalize the knowledge of different groups of scientists. The scientists themselves were determined to share their knowledge with each other. Strife over this issue also flared up in the early 1980s, when the executive branch of the government tried (unsuccessfully) to widen the scope of scientific work which should be classified secret and remain unpublished.

Honesty

Obviously, truthfulness is held to be a virtue quite generally, not just in the sciences. But honesty does have a special place as a core value in the sciences, which is not always typical of other human affairs. We are not always shocked when politicians, lawyers, and businesspeople tell lies. Dishonesty in science, however, is still greeted by some outrage, and rightly so. The reason is not that scientists are considered more virtuous than other professionals, but because honesty as a virtue is more important in doing science. You can engage in a real estate transaction or make a political deal without assuming that the other party is telling the truth; you can't do science, however, without assuming that the other people engaged in the same work are giving you honest information. There is certainly some fraud occurring in science, as evidenced by a number of well-publicized cases. There is also a disturbing trend for some scientists in a few disciplines to be considered "experts" and thereby qualify for large payments in legal proceedings (which can be a corrupting influence). Fortunately, these things are still a very small part of the entire scientific enterprise. Even so, they are a matter of grave concern in the scientific community. Why? Precisely because honesty is considered such an important scientific value.

Curiosity

Curiosity, in this context, is the desire to know more and to better understand nature. In other words, scientists always consider learning more about nature to be a positive good. Such curiosity is not only a part of the personality structure of most scientists, but it's also taken to be one of the values of the scientific community as a whole. Unlike honesty, curiosity is not always considered a virtue by all members of our society. Novel ways of thinking sometimes contradict traditional understandings. Certain religious groups find curiosity a threat to their dogmatic beliefs. One of the staples of science fiction is the scientist who seeks knowledge that humans aren't meant to know. And, of course, there is the adage, curiosity killed the cat. Although curiosity is one of the core values of science, individual scientists might have a conflict between this value and the other values they hold. As a clear and simple example, consider the conflict between my curiosity about how much pain a human can withstand and my ethical revulsion over an experiment to find out. A more subtle and difficult example is the following: Suppose a line of investigation driven by curiosity (and therefore good to perform) gives us knowledge that leads to a new technology we know is very dangerous and/or harmful. Should we undertake this investigation?

Open-mindedness

A scientific result or idea must ultimately be based on evidence, that is, observation and experiment. If accumulating evidence contradicts one of our beliefs, no matter how strongly held, then we must give up the belief. This willingness to change your mind based on evidence is also one of the basic values of science. Of course, there are stubborn scientists who are hard to convince; scientists are humans. It's fair to say, however, that anyone who doesn't share this value of open-minded willingness to alter a belief in response to evidence is not a scientist. Once again, not everyone in our culture shares this value. Generally speaking, the people who don't like curiosity are also not thrilled by open-mindedness. Also, politicians who change their positions based on evidence are sometimes accused of being weak and vacillating (waffling).

Values Employed in Theory Choice

In determining whether one theory is better than another, scientists employ criteria that, from a philosophical point of view, can be considered as values. For example, we generally prefer theories that have greater accuracy, better consistency with other theories, a broader scope of application, a higher degree of simplicity, and are more likely to lead to progress. While these things are values (strictly speaking), they are not really in the same category as the rest of the subject matter in this chapter. Criteria for theory selection are discussed more extensively in chapter 14.

Value-free Science?

You often hear or read the statement that science is objective and/or value-free, but the meaning of this statement isn't always clear. If the statement refers to the actual content of science (e.g., experimental and theoretical results), then the statement is problematic. Arguments over this issue are sophisticated and difficult; fortunately, we don't need to deal with these arguments here because they are unrelated to the present issues of interest (see chapter 15). If the statement "science is value-free" refers to the overall context within which science is done, however, then this statement is utter nonsense. To the extent that science is the activity engaged in by scientists, then clearly science is tied to a set of values, namely, the shared values of the scientific community (which have been the main subject matter of this section).

§2. THE IMPACT OF SCIENCE ON VALUES

We have focused our attention so far on the shared values of scientists, but this is only a small part of the story. Scientific results often lead to new technologies, profoundly affecting human society in ways that can be either useful or destructive. In this way, science becomes entangled in questions of values and ethics that lie far outside the original scope of scientific values.

For example, questions about the behavior of atomic nuclei, which were posed and answered by physicists many years ago, simply involved curiosity about the workings of nature. An unexpected result of the knowledge gained about nuclei was the development of the atomic bomb, a weapon of mass destruction. Similarly, studies of how organisms transmit inherited characteristics, motivated by scientific curiosity, have resulted in the technology of genetic engineering with its accompanying ethical dilemmas. These examples clearly illustrate how seemingly value-free scientific issues quickly become value-laden when the science bears technological fruit. Even in the absence of new technologies, new science can sometimes have unforeseen effects on cultural issues well beyond the scope of the science itself. Increasing knowledge alone can affect the way we think about our values. Let's explore some of these issues (concerning both technology and science) in more depth.

New Technologies

Many of the important issues can be illustrated by looking at a specific case: molecular biology and genetics. Rather than looking at future scenarios of human clones, let's consider the implications of some procedures that are present-day realities. A number of diseases are inheritable, such as Huntington's disease, phenylketonuria, and sickle cell anemia. People with family histories of such genetic diseases are obviously at higher risk for having them. Until recently, however, there was no way to know whether an at-risk individual actually had the condition until the onset of symptoms (which may occur late in life, as in Huntington's disease for example). Scientific advances in our knowledge of genetics now allow us to identify particular genes (or groups of genes) as being responsible for some of these diseases, so we now have the technological ability to tell someone whether or not a condition exists long before any symptoms become evident. Is this a good thing? The ability to determine whether an individual has a genetic disease opens up a range of ethical dilemmas. Unless we have a treatment for the condition, the person might well prefer not knowing. If a doctor knows, is it ethical to withhold the information from the person (or, for that matter, to not withhold it)? If an insurance company finds out, it might refuse to issue health or life insurance; how should society handle that problem? If the person is a minor, who should determine whether a test is performed, the person or the parents? If the test can be performed prenatally (which is often the case), the issues become even more complex due to the possibility of abortion if the genetic condition is detected. Many examples of this sort can be found in the biomedical field. New techniques, for example, now allow us to prolong life far longer than ever before. Few people would argue that this is not good, but even this generally positive outcome of scientific progress opens up some difficult questions. At what point is it proper to allow a suffering and terminally ill person to die? How much of society's scarce resources should be used for the expensive process of keeping very old and very ill people alive, as opposed to improving health care for babies and young children? The issues involved in such questions go far beyond the science and technology that provide the new techniques.

Biomedical applications are not the only technologies that raise issues of values and ethics. In chemistry, for example, the technological goal is often to produce some useful new substance. These substances can be powerful agents to benefit society, but they can also do unintended harm. Chemists have created new pesticides and fertilizers, allowing highly improved agricultural yields. Feeding more people is certainly good. The gains have been accompanied by serious side effects, however, such as environmental pollution and toxic effects on humans and wildlife. Weighing the positive and negative effects on humanity in a case like this depends at least in part on our values as well as our scientific knowledge.

The application of science to military technology is also an area rife with ethical questions. Although there are many examples of military applications, perhaps the most famous is the development of the atomic bomb (largely by physicists). The invention of nuclear weapons, with their unprecedented destructive capability, led many scientists in the postwar era to consider the ethical dimensions of scientific work more seriously than previous generations had done. The fundamental scientific work done earlier in the century to understand the nature of radioactivity and nuclear forces gave no hint of these devastating technological applications; many scientists completely dismissed the possibility of using the nucleus as a source of great energy. By the time they realized that such applications were practical, many scientists in the Allied countries perceived themselves to be in a deadly race with Nazi Germany to produce a bomb. On that basis, they believed their actions to be quite justified ethically. Subsequent events, including the use of atomic bombs on civilian targets and the arms race with the Soviet Union, caused some scientists to regret the decision to build the bomb (some withdrew from military work). Others considered their actions well justified by the Communist threat, and they continued to devise ever-more-destructive weapons.

As a final example, let's take a look at computer science and cybernetics. We again have a case of powerful technologies that can greatly influence people's lives. Realization came early that automated control systems could displace human workers for many tasks, and this trend has continued. Many commentators have noted the massive restructuring of our economy brought on by the information age, and with it the potential for social dislocations and further marginalization of some segments of society. In addition, important issues of individual privacy, censorship, and intellectual property rights have been created by the advance of computer technology and networking (for example, private information about yourself, which you give voluntarily to one organization, will almost surely end up in many other databases without your knowledge or consent). Once again, we are forced to confront new challenges to our old values by the advance of technology. My main point here is that all of the issues we've been discussing exist because new scientific understanding gave us new abilities. I am certainly not arguing that scientific advances are bad because they may create new ethical issues. The gist of all these examples can be summarized by a statement that has been said so many times as to become trite: Science and technology give us power but not wisdom. Any tool can be used for good or for ill, and only we (as individuals and as a society) can choose.

New Science

Science and technology are very different. The previous examples have all involved technology, and concerned science only indirectly insofar as the technological advances were enabled by scientific understanding (see chapter 10). Does science per se have any direct impact on our values? Historically, the answer is certainly yes, as the following examples will illustrate. Whether science must necessarily affect our values is a philosophical question that is difficult to answer (and not really our main concern here).

An early example of the effects of science on values is the impact stemming from the rise of the heliocentric theory. Looked at from a purely practical point of view, it really doesn't make much difference whether we believe the earth or the sun is at the center of the universe. But in the cultural milieu of late medieval Europe, the central location of the earth in astronomy was inextricably associated with the central importance of humankind itself in the grand scheme of things. The central position of the earth could not be dislodged without having people's understanding of their own nature severely shocked. A people's self-image is surely intimately connected with their values. As the debate over heliocentrism unfolded, academic questions of astronomy became increasingly bound up with the more general struggle between the forces of progress and reaction in Europe.

A similar process occurred with even greater intensity in the debate over Darwin's ideas on evolution by natural selection. In this case, the stakes were even higher for the human self-image because humans themselves are part of the subject matter of the theory. Evolution by natural selection effectively removes a barrier between humans and the rest of the animal kingdom; the potential impact of evolutionary thought on values was profound. In the hands of careless thinkers, evolutionary theory led to the perfidious doctrine of social Darwinism (this was an attempt to use scientific advances improperly to justify existing social class inequities). The same strain of thought also contributed to the infamous eugenics movements of the early twentieth century (eugenics is the attempt to "breed" better humans by not allowing the "unfit" to reproduce). These unfortunate consequences were in no way implied by Darwinian thought, but they do illustrate how much potential impact science can have on values, and for that reason how careful we must be when making such interpretations.

Physics has also had an effect on humanity's view of itself and thereby on our values. The equations of Newton's classical mechanics are deterministic. In other words, if we know all the forces on a particle, and if we know its present motion, then we can exactly predict all of its subsequent motions. Actually, such a program was never really possible in practice (see chapter 17), but this practical difficulty didn't affect the conclusions drawn by various eminent thinkers. The idea of a clockwork universe, in which the future is predetermined and free will doesn't exist, became deeply embedded in our culture. The revolutionary new theories of twentieth-century physics (relativity and quantum mechanics) introduced a variety of radical ideas. One of these new ideas was an element of nondeterminism in microscopic events. In other words, we cannot always predict exactly what will happen to a particle even if we know everything that we can know about that particle. Once again, a number of eminent thinkers seized upon this result of physical theory to claim a place for free will in the world, and this idea has also passed into our popular culture.

Our final example deals with issues in which the impact of science on values can be extreme: the study of human beings by modern biology and psychology. A generation ago, the reigning paradigm in psychology was behaviorism and operant conditioning. Some extreme behaviorists claimed that all human behaviors were predictable based on past and present sensory stimuli; they furthermore claimed that mental processing, not expressed in behaviors, was an epiphenomenon of no importance. These ideas imply that free will and reflective self-consciousness are merely illusions, a claim that very strongly influences our values. The behaviorist fad has now passed, only to be replaced by its polar opposite: the claim that who we are as humans is primarily the result of evolution and genetic predisposition. The extreme viewpoint here is that only the genes are of any importance, and that we humans are merely convenient methods for the genes to propagate themselves (another claim that obviously influences our values profoundly). Remarkably, all of these claims have been passed off as purely scientific statements without any value-laden components, a point that we will reconsider near the end of the next section.

§3. THE IMPACT OF VALUES ON SCIENCE

Do the values held by a culture or an individual affect the way science is done, or even the results of science? Whether the actual results (theories and facts) of science are much affected by the values (and other presuppositions) of a culture or individual is hotly debated in academia. We'll examine this issue only briefly here (a more detailed treatment is found in chapter 15). But the more limited question of whether our values affect science in any way at all seems to me to have a clear answer: yes. For example, our cultural values regarding material prosperity influence the amount of effort we spend studying scientific questions that we think will contribute to our prosperity. The values of an individual concerning patriotism and/or pacifism might influence that person's decision whether to work on a scientific project with clear military implications.

The Limits of Scientific Inquiry

Let's look at another example in somewhat more detail: the issue of whether there should be any limits to our choices of which scientific questions to study (and if so, what those limits are). One dimension of this question concerns ethical issues surrounding the implementation of a research study, rather than the scientific knowledge resulting from the research. In this case, the ethical issues may extend far outside the boundaries of science, and there are clear limits to the extent of the inquiry allowed. An infamous example is the Tuskegee study of the long-term effects of syphilis in a group of black men from 1932 to 1972 (the men were purposely not treated, in order to observe the effects of the disease); a stark reminder that freedom of scientific inquiry has sharp limits when overriding ethical principles are involved. Other examples in this category (but where the issues are more complex, controversial, and difficult to resolve) include the use of animal subjects in research and the use of fetal tissue in research.

A much different category involves the question of whether there should be limits on scientific research if that research will lead (or might lead) to dangerous technologies. Restricting the early nuclear physics research that eventually enabled nuclear weaponry might be an example, or the molecular biology research that enabled genetic engineering. We can make philosophical arguments both against and in favor of such limitations, but I think it's a waste of effort to weigh such arguments because in practice the issue is moot: we can't possibly predict all the technological consequences at the time the research is being done. The real issues concern how to control the technologies after they become apparent. These issues are important, but they are not issues about the limits of scientific inquiry. An interesting exception to this point might be research that is undertaken primarily in the hope and expectation that it will lead to a certain technology (for example, plasma physics/fusion power or geriatrics/increased longevity); in those cases, the desirability of the research might well be debated in terms of the desirability of the anticipated technological outcome.

Finally, we can ask the question of whether there should be limits to scientific inquiry because the knowledge produced is something we don't want to know (or shouldn't know). The knowledge might be antithetical to some cherished traditional belief, for example, or might exacerbate tensions between ethnic groups in society. We see here a clash between the inherent values of science (curiosity, open-mindedness) and other values esteemed by our society (or at least by some members of it). A discussion of such deep and complicated issues is beyond our ambitions here, but I will offer an opinion and a comment. My opinion is that freedom of inquiry is not something to be given up lightly, but rather should be considered one of the core values of our culture in general, not just in science. Ignorance is far more likely to harm us than to help us. My comment is that good, solid scientific knowledge rarely causes any problems. The problems usually arise when highly value-laden implications are drawn from scientific results, a procedure that we should always scrutinize as critically as possible.

Risk Assessment

Values also have an impact on science in the assessment of potential risks to society from toxic pollutants, disasters, and so on. The degree to which value-neutral science can be done in this field is controversial, and the question is obviously a significant one. Let's look at the issues more closely. We'll first dismiss from further consideration the extreme cases of ideological zealotry and commercial greed. People who will say virtually anything with no regard for evidence (tobacco industry employees claiming that heavy smoking does no harm, for example) may sometimes be labeled as scientists, but using the word doesn't make it so. These extreme cases have little to do with real science and are not of interest here.

The interesting question is whether legitimate scientific risk assessments can be kept free of presuppositions based on values. A number of people, including many scientists and engineers, believe that this is possible. "Facts are facts, and the results of properly made measurements don't depend on our values." According to this view, we employ our values when we make policy decisions based on the risk assessments; objective science tells us what the risks are, and our values tell us whether these risks are worth taking for the benefits involved. This point of view certainly has some merit, but it has been criticized as an overly idealized picture of the way the process actually works. The flaw in that picture, according to its critics, is that we rarely have a complete and definitive set of data. In fact, the typical situation is just the opposite: only a small amount of very sketchy information is available. Moreover, the information that does exist is often not directly applicable. For example, toxicity studies involving animals and high doses must be interpreted for the case of humans and low doses. No one knows how the dose extrapolations should be done, or what the differences are between the toxic substance's effects on humans and on animals. In trying to formulate a risk assessment based on such inadequate information and high degrees of uncertainty, we are forced to make scientific judgments of various sorts.

Our scientific judgments, say the critics, can easily be influenced by our value judgments in general, especially given the fact that our assessments have importance for real people's lives. For example, a person who generally thinks that industry is overly regulated might well emphasize the following point: large uncertainties allow the possibility that a substance is not very dangerous. A person who generally thinks that exposure to environmental toxins is a major public health problem, on the other hand, might instead emphasize the opposite point: these large uncertainties allow the possibility that the toxic substance is far more dangerous than we can presently document. Neither of these examples is bad science. In both cases, the people did as well as they could with the information at hand; the point is that the large uncertainties in the information allowed their values to influence their interpretation. This influence may or may not be inevitable. But to the extent that it happens, we'll improve the quality and worth of our risk assessments by critically scrutinizing the risk assessment process in terms of both science and values.

Studying Humans, Revisited

We've already looked briefly at the impact of science on values in the scientific study of human beings. But perhaps that influence goes both ways; presupposed values might also affect the science being done in this case. In hindsight, for example, it's quite obvious that much of the work on human heredity (closely related to the eugenics movement) in the early 1900s was tainted by class and ethnic prejudice. There have been at least three major attempts to draw conclusions about humans based on studies of animals: behaviorism, ethology, and sociobiology. In each of these three cases, the claims concerning human nature have gone well beyond the legitimate scientific knowledge. If such claims about human nature are not justified by the purely scientific information available, they must surely contain significant value-laden elements. (I will admit that my judgment here might be controversial, though I don't see how to avoid it.) Moreover, the very claim that such animal studies even *can* tell us something of importance about human nature is a presupposition (which is not to say that it's wrong) rather than an empirical fact. I don't think it's stretching too far to say that this presupposition stems from the values of these investigators.

So far, we've only been talking about the influence of values on the overblown claims made in the name of science in these three cases. We still haven't addressed the question of whether values influenced the scientific methodologies themselves, or the response of the scientific community to these studies. This question is more difficult, and I don't think any definitive answers are possible. It does seem interesting that behaviorism became popular in the United States during the middle of the twentieth century, a time when the general thinking in the culture was receptive to the idea that people are primarily molded by their social conditions. In contrast, ethology grew out of a central European intellectual tradition that existed in cultural and political conditions of instability and authoritarianism; these conditions may have been more hospitable to an emphasis on inborn traits and instincts. Reading too much into such observations is dangerous, however, and any thoughts about the influence of cultural values on these sciences must remain speculative in the absence of a detailed scholarly discussion.

§4. WHERE SCIENCE AND VALUES MEET

What is the proper relationship between science and values? As we've seen, one possible answer is this: "There is no relationship, because science tells us what is, and values tell us what ought to be." But the many examples we've discussed make it clear that this answer doesn't hold up very well under close scrutiny. The sentiment expressed is not completely wrong, just inadequate to cover all of the many situations possible. At the opposite extreme, we see claims that science and values are not only related but that the relationship is hierarchical (i.e., one thing totally controls the other). "Values are created by the brain, which can be studied scientifically; values are just a branch of biology." So say the extreme reductionists. Of course, the extreme postmodernists say something different: "Sciences are always based on unstated presuppositions, which are really value preferences; scientific results are merely expressions of our social values." Such extreme views surely don't do justice to the subtle complexities involved in these questions.

So values and science are related to each other, but the relationship is not hierarchical. Each has an important role to play in examining issues, and these roles are complementary. Science can provide empirical information and a depth of understanding to inform our debates over values and our ethical decisions. But the resolution of these debates and decisions must ultimately depend on a sagacity that comes from outside the realm of science. Because science (through its associated technologies) also provides us with a great deal of power, the issues require careful thought. Both scientists as individuals and also society in general have responsibilities in the consideration of these issues. Some responsibilities of scientists are to exhibit extreme honesty concerning the uncertainties in our knowledge and to carefully scrutinize the broader ramifications of their work. Society as a whole must be involved in making decisions about those issues that cut across the boundaries of science, values, and ethics; the obligation of the scientist (as someone with special expertise) is to join the debate as an informed citizen.

FOR FURTHER READING

The New Scientist, edited by P. C. Obler and H. A. Estrin, Doubleday, 1962.
Philosophy and Science, edited by F. W. Mosedale, Prentice-Hall, 1979.
Between Science and Values, by Loren R. Graham, Columbia University Press, 1981.
Acceptable Evidence, edited by D. G. Mayo and R. D. Hollander, Oxford University Press, 1991.
Science, Technology, and Society, by Robert E. McGinn, Prentice-Hall, 1991.
The Code of Codes, edited by D. J. Kevles and L. Hood, Harvard University Press, 1992.

EPIGRAPH REFERENCES: E. A. Burtt, in *The New Scientist*, edited by P. C. Obler and H. A. Estrin, Doubleday, 1962, p. 259. C. P. Snow, in *The New Scientist*, p. 129.

Chapter 12

QUESTIONS OF AUTHENTICITY: SCIENCE,

PSEUDOSCIENCE, AND HOW TO TELL

THE DIFFERENCE

> The first criterion that a contribution to science must fulfill
> in order to be accepted is a sufficient degree of plausibility.
> Scientific publications are continuously beset by cranks,
> frauds and bunglers whose contributions must be rejected
> if journals are not to be swamped by them.
> *(Michael Polanyi)*
>
> If you go into that realm without the sword of reason,
> you will lose your mind.
> *(Robert Anton Wilson)*

THE PREFIX "pseudo" comes from a Greek word meaning false, so pseudoscience literally means false science. "Pseudo" also carries an implication of counterfeit or deceptive, making pseudoscience not only false science but also false science that pretends to be real. These simple definitions, however, don't really tell us much. We may know that pseudoscience is false science, but how do we know whether some particular body of knowledge or set of claims is pseudoscience or genuine science? What criteria do we have to make this determination? No official set of specific standards has been universally agreed upon to distinguish pseudoscience from genuine science, but there is broad overall agreement among scientists and philosophers on the general principles involved. A number of authors have attempted, with varying degrees of success, to specify a set of criteria. The attitude of many working scientists, however, is that no definition of pseudoscience is needed, because they know it when they see it. Unfortunately, this attitude doesn't help the average person make such a judgment. Also, this attitude reinforces the tendency to label anything you don't like as "pseudoscience," causing the word to lose any precise meaning at all.

In §1, we'll look at a set of characteristics that can be used to identify pseudoscience and distinguish it from genuine science. As I said, these criteria may not be universally accepted, but they are clear, sensible, and useful; I doubt whether many working scientists (or philosophers of science) will have any very serious objections to them. Following the general discussion of the defining characteristics of pseudoscience, we'll examine several specific examples of pseudoscience in detail. In each of the examples, we'll discuss exactly how the criteria apply to that particular case and show why it qualifies as pseudoscience (i.e., why it does *not* qualify as genuine science). The three examples we will look at are the work of Immanuel Velikovsky, perpetual motion machines, and creation science.

§1. Defining Characteristics

Static or Randomly Changing Ideas

One of the hallmarks of real science is growth and progress in our understanding. Ideas change over time as new discoveries are made; novel research fields open up as new techniques become available and new questions are asked; and fragmentary facts become integrated into coherent theoretical overviews as the knowledge base increases and creative minds work to comprehend this knowledge. Old ideas and knowledge are not discarded in this process; instead, they are reinterpreted in light of the new understanding that has been achieved. Many examples of this progressive growth in scientific understanding are chronicled in Part I. In contrast, the ideas in a pseudoscience either remain static or else change randomly. Either way, there's no discernible progress. There is a good reason that we see no progress: the pseudoscience has neither an anchor in a well-established foundational body of knowledge, nor any systematic comparison with observation. If there is some dogmatic idea behind the pseudoscience, this idea remains static, since there is nothing to change it. If not, then ideas come and go at random because there is no particular reason to accept some and reject others.

Vague Mechanisms to Acquire Understanding

This brings us to our second criterion. In genuine science, the goal of the activity is to achieve some coherent understanding of our observations. We must reject our understanding if it is incoherent or if it conflicts with observations and experiments (I'm leaving out a few subtle points here; see chapter 14). In other words, there are certain general procedures that virtually all scientists would agree are valid, even if the details of how these procedures are applied may not be identical in every particular case. Most of this book is concerned with these valid procedures and how they are applied in a number of interesting examples, so I won't elaborate here. The present chapter is concerned instead with pseudoscience, in which the procedures are only caricatures of those found in genuine science.

Understanding in pseudoscience might be based on many different premises, which may be neither coherent nor consistent with observation. For example, understanding might be based on a great idea that seems to explain everything but that is so vague (and so vaguely connected to anything else) that the idea has no actual content. Or understanding might be based on connections between things that are not logical deductions or empirical findings, but rather just imagined connections in the mind of the pseudoscientist (between colors and musical scales, for example, or between sunspots and business cycles). Sometimes, two or three observations are used to prop up a vast array of speculative thought, none of which is related to or supported by any other observations. Or, to give one last example, the premises of a pseudoscience are often "proven correct" by (allegedly) proving some other premise wrong (e.g., this must be a sparrow because I've shown you it's not a pigeon). This example starts to overlap with our next criterion.

Loosely Connected Thoughts

Rigorous logic, a strict chain of deductive reasoning with no gaps or weak spots, is highly prized in the sciences. This ideal is sometimes not possible: there are gray areas and matters of interpretation in real science; creative new work may depend on intuitions and metaphors; and, of course, scientists occasionally make mistakes. Nevertheless, the ideal of rigorous logic is still maintained as something to strive for. If a scientist makes an error of logic, the legitimate task of other scientists is to find this mistake and point it out. In pseudoscience, on the other hand, we often find wide, gaping holes in the logic; indeed, we often find that there is no logic at all, just some loosely connected thoughts. For example, a man named Ignatius Donnelly published a book in 1882 claiming that the legendary continent of Atlantis existed. He based his claims on the similarities between the ancient cultures of Egypt and South America (pyramid building, flood myths, embalming, etc.). His argument was that these similarities could only be explained by the existence of an earlier Atlantean culture, situated geographically between these two areas, which colonized them both. In this fairly simple case, the low quality of the logic is apparent. Unfortunately, promoters of pseudoscience often use technical-sounding words and scientific jargon, making it more difficult (though not impossible) for people without scientific backgrounds to spot the lack of meaning in the way words are used. We'll see several examples of both obvious and not-so-obvious logical flaws in the case studies. A point to emphasize is that when logical errors do occur in science, there is a way to correct them built into the normal process of doing science; in pseudoscience, such correction processes don't exist.

Lack of Organized Skepticism

This last point brings us to our fourth criterion: organized skepticism. A new idea or result in science is usually presumed wrong until it is shown to be right. The typical way to present results to the scientific community is to publish the results in a professional journal. But before it can be published, new work must undergo peer review, which means that it's sent to other scientists for criticism and judgment; only work judged as worthwhile will be published (see chapter 10). The norm in science is to subject ideas, experiments, and interpretations to criticism in order to weed out bogus results. The results that survive this process become a well-established consensus, and new results that contradict this consensus are greeted by particularly severe skepticism. On the other hand, even the consensus remains subject to criticism, and that criticism becomes severe if new and contradictory results (having survived their own skeptical scrutiny) start to accumulate. Oddly enough, skepticism keeps open the possibility of change even as it tends at the same time to foster conservatism in science.

No such tradition of organized skepticism is found in pseudosciences. For those pseudosciences that are based on a preconceived belief, skepticism is in fact forbidden. For some promoters of pseudoscience, selective skepticism of other bodies of knowledge (including mainstream science) is practiced, but not of their own. For many of the believers of various pseudosciences, though, skepticism is merely an irrelevant concept. They simply don't engage routinely in any practice of critical thinking. Needless to say, the skepticism with which scientists greet pseudoscience is generally unwelcome. This skepticism is interpreted as the close-minded response of someone invested in protecting an orthodox status quo. People who make this interpretation don't realize the important role of skepticism in scientific thought.

Disregard of Established Results

Our last criterion is not only misunderstood by promoters of pseudoscience, its importance is also underestimated by many people who evaluate the competing claims of a science and a pseudoscience. Scientific advance virtually always builds on previous work (as Newton phrased it, he "stood on the shoulders of giants"). We see this time after time in the stories of discovery related in Part I. Even revolutionary changes arise out of a well-understood context, and such changes always account for the tenets of the outmoded viewpoint. Well-established results have become so through a long hard process of critical scrutiny, and these results remain established because they continue to explain a wide variety of observations and experiments in a coherent and satisfying manner. For all of these reasons, scientists work from a sturdy foundation of accepted ideas even as they try to extend our knowledge into new and unfamiliar areas.

In pseudoscience, on the other hand, we find a rather cavalier disregard for established results. Indeed, contradicting known results is often taken to be a great virtue because it shows how new and exciting the ideas are. Rejecting the ideas of a pseudoscience because they conflict with everything else we know is considered (by the believers in the pseudoscience) to be close-minded and authoritarian. While I agree that any idea, no matter how unorthodox, initially deserves an open-minded hearing, this does not imply that the unorthodox idea and the established idea should be considered on an equal footing. For the reasons I've outlined, the unorthodox idea must be subjected to a much greater burden of proof. Those who engage in pseudoscience don't accept the obligation to provide such proof, or even to take into consideration the foundational knowledge that has been developed by the sciences over hundreds of years. Perhaps one reason that people who work in a pseudoscience feel free to ignore established results is that such people mostly work in isolation from any broader intellectual community. Whereas a scientist works to integrate results into a larger framework (which ultimately includes all of the sciences), the pseudoscientist works alone (or in a self-contained group that maintains no intellectual contact with anyone outside the pseudoscience). This isolation is actually yet another criterion by which to identify pseudoscience.

Some Afterthoughts

Not every activity that meets some or all of these criteria is necessarily a pseudoscience. Only those activities that meet the criteria *and also claim to be sciences* are pseudosciences. For example, loosely connected thoughts not ordered by logic are perfectly appropriate (and may be quite profound) in poetry. Ideas that remain static and unchanging over thousands of years are found, again very appropriately, in religions based on scriptures. Only a fool would call poetry and religion pseudosciences. A more subtle and interesting example is provided by alchemy, which is very often labeled as pseudoscience. As cultural historians such as Mircea Eliade and Titus Burckhardt have shown, however, traditional alchemy was a method of spiritual initiation rather than a misguided attempt to do science as we know it; in that sense, alchemy is not actually a pseudoscience. To reiterate, a pseudoscience is an activity that claims to be a science but is not a genuine science based on the criteria listed above.

Our criteria are somewhat formal, and don't include personal characteristics that are sometimes associated with practitioners of either pseudoscience or science. For example, pseudoscientists may have a tendency to feel persecuted when their ideas are rejected, harbor a seemingly personal antipathy toward mainstream science, or have a hidden agenda (such as making money or promoting a religious doctrine). I'm mentioning these personal traits because they do sometimes occur, but they are not defining characteristics of pseudoscience, they don't always occur, and we have only a slight interest in them here. In describing the differences between science and pseudoscience, I have also been considering science in a normative sense, a kind of idealized science. In other words, I have been describing how scientists ought to behave, not necessarily how they always do behave. Real scientists, being human, sometimes fall short of this ideal (for example, scientists have been known to cling stubbornly to an idea in the face of contradictory evidence, or uncritically jump on a bandwagon). But this truism is irrelevant. The point here is that pseudoscience doesn't even share the normative ideals of genuine science.

Lastly, note that each of the five criteria listed above has significant overlap with all of the other criteria. Although I have separated them into categories for conceptual simplicity, the criteria are really all interrelated. For example: lack of skepticism allows the presence of loosely connected thoughts to go uncriticised; the lack of mechanisms to acquire understanding stifles the ability to make progress; and so on. While isolated aspects of one or two criteria may occasionally creep into real science, activities or ideas that meet most or all of these criteria (operating together and reinforcing each other) are certainly pseudoscientific.

§2. VELIKOVSKY

In 1950, Immanuel Velikovsky published a book called *Worlds in Collision*. This book was the result of a decade's research into the myths of many ancient cultures, and the major thesis of the book was extraordinary to say the least: cataclysmic events, found in the myths of almost every culture in the world, had their common origin in real disasters caused by collisions between the earth and other members of the solar system. The book was furiously attacked by several members of the mainstream scientific community, but those events are a story of their own, which I'll relate at the end of the section. Our primary interest is whether this work should be considered science or pseudoscience. The first question to ask is whether Velikovsky claims that his work is science. Based on comments in the preface of the book, on explicit comparisons between his ideas and standard scientific ideas, and on numerous statements made later by Velikovsky and his supporters, there is no doubt that he claims his work is scientific. How then does the work compare to our criteria for pseudoscience? Before answering this question, let's take a more detailed look at the content of *Worlds in Collision*.

It's a massive book, 400 pages long with about half a dozen citation footnotes on each page. Velikovsky quotes from legends of flood, fire, earthquake, battles between deities, and so on from many, many cultures (Hebrew, Greek, Hindu, Chinese, Egyptian, Assyrian, Mayan, Choctaw, Japanese, Babylonian, Samoan, Persian, Finnish, Eskimo, Ovaherero, etc.). Most of the book is taken up by quotations from his sources, detailing these legends and myths. The Hebrew accounts in the Old Testament of the Bible are particularly prominent. To account for the similarities in the myths of so many cultures, Velikovsky believes that some real global catastrophes must have occurred. Based on the details he has found in his historical studies, Velikovsky concludes that the following events happened: A large planet-sized object was ejected from Jupiter and became a comet. This comet passed close to Earth, causing the plagues of Egypt during Exodus, along with other disasters recorded in other places. Earth passed through the tail of the comet. This tail was composed of hydrocarbons, which rained down in the form of petroleum (rains of fire) and carbohydrates (manna from heaven). The comet left Earth, went around the sun, and came back in 52 years to temporarily stop the rotation of Earth as related in the Book of Joshua. The comet later passed near Mars, causing Mars to leave its orbit and pass near Earth a few times, causing (among other things) the destruction of Sennacherib's Assyrian army. The collision of the comet and Mars also inspired a number of passages in the *Iliad*. The eventual result of this collision was that Mars entered its present orbit, and the comet became the planet Venus (in its present orbit).

This highly abridged account leaves out a lot, but we have enough material now to begin a scrutiny based on our pseudoscience criteria. The first thing that you may notice in this scenario is the total disregard for classical mechanics, which was exceedingly well established and had been for some time (see chapter 4). For example, the amount of energy needed to eject Venus from Jupiter is impossibly high; it's virtually impossible for Mars to have turned the elliptical orbit of Venus/comet into the almost circular orbit of Venus; conservation of momentum is violated by some of these actions; and so on. Velikovsky claims that his work only contradicts the assumption that gravity is the sole force involved, and that electric and magnetic forces between planets can fix the situation. But he offers no theory or calculations to justify this assertion, and in fact calculations show that such forces (which are also well understood) cannot fix the situation either. In his book, Velikovsky claims that his ideas are merely an unorthodox alternative to unproved assumptions about the stability of the solar system, a claim that many readers accepted. But his work actually contradicts a great deal of science that we are very confident is correct. Velikovsky's bland refusal to worry about these contradictions is a sign of pseudoscience.

Now, contradicting established results doesn't by itself make his work wrong, but this does increase the burden of proof to show that the work is right. How does his work measure up to this increased burden of proof? This is an interesting question, because the book certainly offers a large amount of evidence. Page after page of quotations from culture after culture are given on each topic (example: east becoming west and west becoming east). From all this, Velikovsky concludes that some physical event happened (example: the direction of Earth's rotation changed). The cumulative effect of all these similar legends is indeed impressive. But the evidence is not critically examined. The reliability and authenticity of the sources isn't assessed. We don't know what date each of the sources refers to (they would all have to refer to the same date if they described a real event). We don't know the contexts of the quotes. And no alternative explanations are considered by Velikovsky. There is, in short, a lack of skeptical scrutiny in the work, making the evidence appear better than it really is. This lack of skepticism is another one of our criteria for pseudoscience.

The book is also based almost entirely on loosely connected ideas as opposed to logical deductions. For example, a mythical description of a battle between Ares and Pallas Athena becomes a physical collision between Mars and Venus. The thunderbolts of Zeus become electrical discharges between planets. The hypothetical hydrocarbons in a comet's tail can become either oil or food, depending on the legend being discussed. In mythology, Athena sprang from the head of Zeus; so the planet Venus was ejected from Jupiter. A local myth is first assumed to refer to a real event; the event is then assumed to be a global catastrophe; the global catastrophe is then assumed to have a cosmic origin; the cosmic origin is then assumed to be a particular planet passing by; the planet is then assumed to have some particular complicated history. Despite the length of the book and its wealth of detail, there are never any carefully constructed logical connections made between any of the assumptions in this long chain. Instead, the same assumptions are simply repeated over and over. The evidence upon which Velikovsky bases his conclusions consists almost entirely of myths, legends, and ancient writings, not observations of nature. (A few pages are devoted to geophysical evidence that couldn't be interpreted at the time, contributing little to his work.) Yet, we are asked to base our understanding of nature (the solar system) on this evidence. This is another sign of pseudoscience (namely, our criterion of vague methods to acquire understanding).

Velikovsky's supporters may think this criticism is unfair because he did make several observational predictions (some of them correct). These predictions have been used by many people to argue that Velikovsky was right. Here are some of his predictions: Venus and Mars should be hot, due to collisions, electrical discharges, going near the sun, and so on; Venus should have hydrocarbons in its atmosphere; and Mars should have argon and neon as major components of its atmosphere. Do these predictions blunt the charge that Velikovsky's work is pseudoscience? Venus was eventually discovered to be much hotter than expected by mainstream science in 1950. This discovery has been promoted as a striking confirmation of Velikovsky's work, which is in turn a reason to take his ideas seriously. But this prediction is not part of a coherent picture based on Velikovsky's ideas, because heat generated by events thousands of years ago (which his model requires) would have long since radiated away, leaving cold planets. In contrast, the scientific research that discovered the high temperature of Venus also discovered that Venus has a lot of carbon dioxide in its atmosphere, leading to a large amount of greenhouse warming (see chapter 10). This greenhouse effect (which had actually been suggested speculatively by one scientist even before 1950) accounts (quantitatively) extremely well for the temperature of Venus.

But even if we didn't have a good explanation for a hot Venus, we cannot conclude that Velikovsky is right, because his prediction was based on erroneous logic. Also, his predictions about the temperature and atmosphere of Mars are wrong (a fact usually left unmentioned by his supporters). His prediction of hydrocarbon clouds on Venus was thought to be correct for a while, and highly touted as another striking confirmation. Later work has shown that the clouds are actually made of sulfuric acid. Again, the interesting point is not so much that this prediction is wrong. Even if the prediction were right, the proper conclusion (given the fact that there was no valid logic supporting the prediction) would not be that Velikovsky's ideas are correct. We would merely have a remarkable coincidence. We've seen so far that Velikovsky's work should be classified as pseudoscience based on several of our criteria: disregard of established results, loosely connected ideas, lack of skeptical scrutiny, and vague methods to acquire understanding. Our other criterion was lack of progress and growth. His work qualifies as pseudoscience based on this criterion also, since his ideas cannot be modified based on testing, or suggest new research directions. Instead, we are given a storyline, which we can either accept or reject. Velikovsky was an interesting and imaginative thinker, and he was a patient, thorough collector of ancient myths and legends. But his work, whatever other virtues it may possess, is not science.

The controversy between Velikovsky and mainstream science is an interesting story itself. A number of scientists, particularly astronomers, were livid with rage that *Worlds in Collision* had been published. They attacked the work in public (reviews, letters to the editor, etc.), and they threatened the publisher with a boycott of its textbooks unless it stopped publishing the book. This censorship, combined with the abusive and disdainful treatment of Velikovsky and his work, has become an issue of its own, separate from the merits or flaws of the book. The behavior of the scientists was not well received in many quarters, being interpreted as an illegitimate attempt to suppress new and rival ideas. By making Velikovsky into a martyr, the scientists probably increased his support in the general public (presumably the opposite of what they wanted to do) and hardened the position of his followers. The inappropriate behavior of Velikovsky's detractors, however, doesn't improve the argument of his book. We don't need to censor pseudoscience; we just need to learn how to recognize it.

§3. Perpetual Motion

A perpetual motion machine is just what it sounds like: a machine that runs forever. More particularly, it's a machine that runs forever without any fuel or source of energy to keep it going. A perpetual motion machine can continue to do useful work for us without needing any supply of gasoline, coal, electricity, and so forth. We get something for nothing. It sounds too good to be true. In fact, it is too good to be true. Perpetual motion machines do not and cannot exist. The impossibility of perpetual motion was already a current idea among the ancient Greek philosophers. Practical folks had also figured out long ago that there are no perpetual motion machines, probably based on the innumerable failed attempts to build one. Simon Stevin, back in 1586, used the nonexistence of perpetual motion as the starting point of his derivation for vector resolution of forces (in other words, Stevin felt he could take it for granted that perpetual motion was absurd). Scientific understanding of the reason why perpetual motion is impossible came later, when the conservation of energy principle was formulated in the nineteenth century. Conservation of energy is now one of the bedrock foundational principles of the sciences (see chapter 17).

Attempts to invent a perpetual motion machine have a long and fascinating history. A typical design might be, for example, a water wheel that runs a pump; the pump delivers the water to a higher elevation; the water runs down, turning the water wheel; and so it goes, forever. Another popular idea uses flywheels with hinged weights on them that are supposed to use gravity to keep the flywheel turning. Many such clever mechanical contraptions and ideas have been proposed, but none of them have ever been known to work. Of course, we wouldn't expect any of them to work based on our present scientific understanding. The U. S. Patent Office does not accept applications for perpetual motion machines anymore; the Paris Academy of Sciences refused to consider claims of perpetual motion back in 1775. But people are still claiming, even now, to have invented perpetual motion machines. Moreover, their claims are taken seriously by journalists, politicians, and investors. How can this be? Are all these people unaware of the history of failure in such attempts, and unaware of fundamental scientific results? They are in fact aware of these things, but are not concerned by them. The inventors believe they have discovered important new scientific principles that can make perpetual motion a reality for the first time in history. What is this new science that has been discovered (so they claim) by the perpetual motion machine inventors? Basically, they are saying that their machine can tap into a vast reservoir of energy that is always present but unnoticed. No violation of the energy conservation principle is involved because a source exists for the energy produced by their machines. Now, one of the modern theories of physics (quantum field theory) does in fact have such a large energy reservoir as one of its concepts (called zero point energy or vacuum fluctuation energy). Invoking such a concept makes their claims sound fairly plausible. Is the idea that we can tap into the vacuum energy to make a perpetual motion machine science or pseudoscience?

The issue has now become subtle for the following reason: One of our criteria for pseudoscience is the disregard of established results. At first, we might say this criterion is obviously met, since perpetual motion is utterly forbidden by conservation of energy. But by using the concept of the vacuum energy, the promoters of perpetual motion can claim that conservation of energy is not violated; we are not creating energy out of nothing, we are simply using some existing energy. Since the vacuum energy qualifies as an established result, our conclusion seems less obvious now. However, we are ignoring a key point: the *same theory that tells us the vacuum energy exists also tells us that we can't use it.* Taking this point into account, we see that the reasoning of the perpetual motion promoters does indeed disregard established results. Surely it's illogical to use a concept in a way that is clearly prohibited by the very structure of the theory in which the concept arose.

Another criterion is the lack of logical connections. To be science, their explanation would need to specify unambiguously the mechanism by which their machine extracts energy from the vacuum. Instead, we get strings of technical-sounding words that don't actually have any meaning (example: "The flow of load current does not create any anti-torque and therefore, the output power is greater than the input power used to rotate the device"). Isolated concepts, which don't fit together in any meaningful way, are used. The perpetual motion machines are real (they believe), and so require an explanation. But the explanation is not sought by doing controlled experiments and trying to interpret the results within some constructive framework tied to the rest of science. Instead, a pseudo-explanation is created out of whole cloth, to somehow account for the very existence of the machine. Vague pictorial analogies are employed in place of logical connections between empirical observations ("I use the analogy of a tornado. A tornado concentrates a force in a small volume, a force that was always present, it's just changed form."). There is no attempt to predict quantitatively the output of the machine in response to varying conditions. All that we have are some scientific-sounding bits and pieces (a technological device, some fancy terminology, a quantum field theory concept), but none of it hangs together. The pieces don't fit together into a picture. And this is why it's not science.

Another sign of pseudoscience is the lack of growth and progress in the field. Perpetual motion promoters actually use this argument against mainstream science; they claim that conventional scientists are dogmatically attached to outmoded ideas, while they themselves are enlightened prophets of progress. We've already discussed the validity of this argument. The real issue is whether there has been any progress in understanding the behavior of the alleged perpetual motion machines (or the science underlying this behavior). As we've already seen, there hasn't been any such progress because the inventors are not even really trying to achieve any new understanding. Instead, the same pseudo-explanations are simply repeated over and over.

I have not yet addressed one question, which you have undoubtedly been asking yourself. Haven't these machines been tested? Determining whether they work, once and for all, should be fairly easy. I've saved this question to discuss in conjunction with another criterion of pseudoscience, organized skepticism. The inventors of these machines have apparently convinced themselves that their measurements prove they are getting out more energy than they put in. When other scientists and engineers point out the potential flaws in the measurements that might incorrectly lead to that conclusion, they are ignored. When independent professionals have tested the machines with the best possible equipment, the machines have behaved as conventional science predicts they should (i.e., no perpetual motion). The inventors remain unconvinced. Apparently, lack of organized skepticism also qualifies perpetual motion as a pseudoscience. Of course, the simplest test would also be the most convincing: plug the machine into itself and let it run without an external source of energy. The inventors have not done this, but they are sure they could if they wanted to.

§4. CREATION SCIENCE

The evolution of life on earth is one of the central organizing principles of modern biology. Evolution, however, has always been a controversial idea among some groups, including certain fundamentalist sects in the United States. The members of these sects believe that everything written in the Bible is literally true, and their dislike of evolutionary thought is based on the fact that it contradicts part of the Bible. Several decades ago, a number of fundamentalists with some scientific background joined together to invent something called creation science. The basic thesis of creation science is that the universe was created, in its present form including all known species of plants and animals, about five thousand years ago; and that this conclusion is defensible on purely scientific grounds. Because of these claims for scientific validity, we may subject creation science to the kind of critical analysis outlined in §1. Based on our criteria, is it creation science or creation pseudoscience? Let's first clear up any misunderstandings caused by the close relationship between creation science and creationism in general. Creationism need not make any claims for scientific validity. We can legitimately postulate a divine being who created the universe in a way that makes it *appear* as though the universe had a long and involved history before this act of creation. This argument was in fact made in great detail in a book called *Omphalos*, written by Philip Gosse in 1857, in response to mounting geological evidence for a very old earth. The claim is based purely on religious faith; cannot be affected in any way (supported *or* denied) by observations or evidence; and has nothing whatever to do with science (see chapter 9). Whether we accept or reject creationism on other grounds is irrelevant to our present discussion of science and pseudoscience. Creation science, on the other hand, very explicitly claims that observations and evidence lend scientific credibility to the idea of special creation. We can (and I do) legitimately reject this absurd claim.

Our very first criterion for pseudoscience was lack of progress or growth in our understanding. In the case of creation science, the understanding of its practitioners must remain static by definition because they assume up front that special creation occurred. Looking for evidence is an afterthought, engaged in only for the purpose of supporting their preconceived conclusion. This conclusion will not change, regardless of any possible observation or argument put forth; the creation science promoters are quite candid about the static nature of their view on creation. Nothing can change their view, a stance that is totally antithetical to real science but quite characteristic of pseudoscience. Another criterion, closely related to the first one in this case, is the lack of organized skepticism. These two criteria are related here because an unchangeable preconceived idea cannot, by its very nature, be subjected to skeptical scrutiny. Skepticism implies asking yourself whether an idea is correct or not. Concluding that special creation is incorrect, however, has been forbidden under any circumstances by the believers in special creation. Of course, they have plenty of skepticism for the conclusions of mainstream biology, geology, and physics. But that's not scientific skepticism, in the sense of doubting all premises in order to arrive at a correct picture; instead, selected pseudoskepticism is directed toward any ideas that disagree with their preconceived conclusions. Now, since scientists are genuinely skeptical, they criticize each others' ideas. Creation science writers routinely quote scientists' critical comments in an attempt to cast doubt on the validity of evolutionary ideas. Creation science itself is invulnerable to this tactic, for the reasons I've outlined. Ironically, these characteristics of real science (being open to skeptical criticism, and changing ideas) are exploited by creation science writers to attack evolutionary thought, misleading people who don't understand what science actually is or how it works. As we've seen, however, these tactics are based on aspects of creation science that in fact identify it as a pseudoscience.

Looking at the mechanisms used to acquire understanding in creation science, we see that yet another criterion of pseudoscience is met. In a way, no mechanisms are used at all to *acquire* understanding in creation science; there's no need for them because the answers are already known ahead of time. Creation science has never even attempted to construct a positive research program by which our understanding might be increased. Instead, most of their writings are merely attacks on evolutionary thought, employing the dubious logic I mentioned in §1 (i.e., "If something is wrong with evolution, then special creation must be right."). To the extent that creation science does have something positive to offer, it's an attempt to reconcile our observations with the Biblical accounts (assumed to be true). If we rather broadly interpret "acquiring understanding" to include this sort of activity ("How do we understand what we see in a way that doesn't conflict with our beliefs?"), then the mechanisms we find are vague, illogical, and ad hoc. Instead of fitting all the evidence together into a coherent picture, each piece of evidence is considered in isolation from all of the others. For each of these isolated pieces, the practitioners of creation science devise some kind of separate explanation. To assess the quality of these isolated explanations, we must move on to another of our criteria: loosely connected thoughts. Due to the underlying goals of creation science, its arguments don't need to have any real validity. The arguments merely need to sound plausible when not subjected to any critical scrutiny. We are therefore not surprised to find vague statements and loosely connected ideas instead of well-thought-out logical connections. Let's illustrate this point with an example: the creation science explanation for the fossil record found in layers of rock. The fossils (mostly of extinct creatures) certainly exist, and so creation science is obligated to give a scientific explanation of them. The creation science explanation is that the fossils are the preserved bodies of creatures who died in the Great Flood (the one written about in the Bible, in which all life perished except for that on Noah's Ark).

As a scientific explanation of the fossil record, this story has a major problem: it doesn't account for the different types of creatures found in different layers of rock (geological strata). As we go into deeper and deeper layers, the bodies of the creatures become more primitive; only water-dwelling creatures appear below a certain level; and so on. An evolutionary picture accounts for these facts quite naturally, fitting together nicely with the geological picture of the layers as sedimentary rocks deposited over a long period of time. In other words, lower strata are older, and for that reason contain earlier creatures that are more primitive. The creation science picture (simultaneous death by flood) implies that all creatures should be mixed together in all the rocks. To account for the layering, various separate and unconvincing ad hoc explanations need to be devised. The basic idea simply doesn't explain what is observed. Other questions that this idea doesn't answer include the following: Why were water-dwelling creatures killed by a flood? Why aren't there huge numbers of human remains in the fossil record? How did creatures repopulate islands (like Australia) after the flood? And so on. The point here isn't that the creation science explanation of the fossil record is wrong; the point is that this explanation doesn't really explain anything. We can never *prove* that this account is wrong (after all, the bodies might have miraculously sorted themselves). Nor can we *prove* that the evolutionary account is right. As we've emphasized, the role of science is not to provide certainty (which it can't do; see chapter 14) but rather to provide coherence (which it sometimes does stunningly well). The promoters of creation science argue that, because neither evolution nor creation science can be proved, both are equally valid. This argument, which is central to their program, is based on a profound ignorance of what science is.

Our last criterion of pseudoscience is disregard of established results. The advocates of creation science claim that their work is consistent with sciences other than evolutionary biology. Creation science was devised mainly as a tool with which to debunk evolutionary thought. For this reason, promoters of creation science often use isolated fragments of mainstream sciences (e.g., thermodynamics), which they take out of context and apply incorrectly, as part of their efforts to attack evolution. In this way, they try to convey the impression that their ideas (as opposed to evolution) are really consistent with many established results. The impression they convey, however, is false. *The assertion that the earth is only a few thousand years old flatly contradicts the results of many mainstream sciences:* nuclear physics (radioactive decay dating), cosmology (microwave background), astronomy (redshift measurements), geology (plate tectonics, erosion), and so on. In its disregard for the well-established results of the mainstream sciences, creation science easily qualifies as a pseudoscience.

To summarize, creation science is an endeavor that starts with a preconceived idea and distorts evidence to fit this idea. We achieve no increased understanding of our observations of the world because creation science isn't based on those observations. The statements found in creation science are vague, isolated fragments, having no coherence and no logical connections to each other or to observational evidence. Creation science forbids its practitioners to engage in critical scrutiny of its assertions (also forbidding disagreement with the assertions). Creation science is inconsistent with important results in a variety of real sciences. In short, creation science amply satisfies every one of our criteria for a pseudoscience.

FOR FURTHER READING

Worlds in Collision, by Immanuel Velikovsky, Macmillan, 1950.

Fads and Fallacies in the Name of Science, by Martin Gardner, Dover Publications, 1957.

Scientists Confront Velikovsky, edited by Donald Goldsmith, Cornell University Press, 1977.

Abusing Science, by Philip Kitcher, MIT Press, 1982.

Ideas of Science, by B. Dixon and G. Holister, Basil Blackwell, 1984.

In the Beginning . . . A Scientist Shows Why the Creationists Are Wrong, by Chris McGowan, Prometheus Books, 1984.

"Power Structures," by Tom Chalkley, *City Paper* (Baltimore), vol. 14, no. 26 (June 29, 1990), p. 8.

Understanding Science, by Arthur Strahler, Prometheus Books, 1992.

A Beginner's Guide to Scientific Method, by Stephen S. Carey, Wadsworth, 1994.

At the Fringes of Science, by Michael W. Friedlander, Westview Press, 1995.

EPIGRAPH REFERENCES: Michael Polanyi, *Knowing and Being*, University of Chicago Press, 1969, p. 53. Robert Anton Wilson, *Cosmic Trigger*, Simon & Schuster, 1978, p. xix.

Chapter 13

CONTENTIOUS QUESTIONS: THE SHADOWY BORDERLANDS OF SCIENCE

"Then the one called Raltariki is really a demon?" asked Tak.

"Yes—and no," said Yama. "If by 'demon' you mean a malefic, supernatural creature, possessed of great powers, life span, and the ability to temporarily assume virtually any shape—then the answer is no. This is the generally accepted definition, but it is untrue in one respect."

"Oh? And what may that be?"

"It is not a supernatural creature."

"But it is all those other things?"

"Yes."

"Then I fail to see what difference it makes whether it be supernatural or not—so long as it is malefic, possesses great powers and life span and has the ability to change its shape at will."

"Ah, but it makes a great deal of difference, you see. It is the difference between the unknown and the unknowable, between science and fantasy—it is a matter of essence."

(from Lord of Light *by Roger Zelazny)*

SOMETIMES, results are reported that lie far outside the scientific mainstream. These unorthodox results are rejected by most, but not all, scientists. Although cases like this are occasionally labeled as pseudoscience in order to attack their validity, I believe this label is wrong. As we have seen (chapter 12), the defining characteristics of pseudoscience concern the methods of thinking, not the unlikelihood of the content. Pseudoscience seems an inappropriate description of cases where all parties agree on the validity of basic scientific methodology, even if there is heated disagreement over whether it's being applied properly. Another label that has been applied to the subject matter of this chapter is "pathological science." Once again, some less emotional term might be a better choice. But rather than quibbling over terminology, let's look at some cases in detail. Of the many subjects that might be discussed in this chapter, I have chosen two: cold fusion and parapsychology.

§1. COLD FUSION

Nuclear Fusion

The word "fusion" in this context refers to a process in which lighter atomic nuclei combine together (fuse) to form heavier nuclei, releasing energy. Some of the mass of the nuclei is converted into energy during the fusion process. Fusion is the process that drives energy production in the sun and other stars, so fusion is ultimately responsible for almost all of our energy on earth. Fusion is also the energy source in the hydrogen bomb, a terrifying weapon of mass destruction. The name "hydrogen bomb" refers to the fact that hydrogen nuclei fuse together into helium nuclei. (To be more precise, nuclei of the isotope of hydrogen known as deuterium are the nuclei that fuse. A hydrogen nucleus is a single proton, while a deuterium nucleus has both a proton and a neutron.) For many years, scientists and policymakers have been hoping that fusion can someday be used as a new source of energy for our society. The required fuel is abundant (deuterium replaces hydrogen in a certain fraction of the water naturally occurring on earth; such water is called heavy water), and the radioactive waste products are few, making fusion a highly desirable energy source.

Before recounting the cold fusion story, which is fairly recent, let's look at the efforts to harness fusion energy by standard techniques; these efforts, which have been going on for over forty years, provide the context for our story. Fusion occurs naturally in the sun because the sun is very hot. At the fantastically high temperatures found in the sun, the nuclei and electrons of atoms are separated, forming what is known as a plasma. The nuclei are positive and the electrons are negative, so the plasma is a kind of charged particle gas. The charged particles making up the plasma in the sun have extremely high energies and speeds because of the high temperatures. These high energies and speeds enable the colliding nuclei to get close enough to each other to fuse, overcoming the mutual repulsions caused by their charges (which ordinarily keep them apart, preventing fusion). Along with the extreme temperatures needed, the density of the plasma must also be high enough to yield significant numbers of collisions. If the density can be kept high enough after the fusion reaction has started, this reaction will be self-sustaining because the energy released by the fusion process will keep the temperature high. The sun's own gravity keeps the density high enough in the sun. In a hydrogen bomb, the density doesn't need to remain high very long, because the explosion lasts only a short time (the needed high temperature is provided by an atomic bomb explosion as a trigger). The challenge to the plasma physicists has been to create a plasma with a steady high temperature and sufficient density so that a controlled self-sustaining fusion reaction can be maintained on earth. If this can be done, we might be able to use nuclear fusion to satisfy the world's need for energy.

The problem with creating these conditions may already be apparent to you. At temperatures like those in the sun, any container in which we tried to keep the plasma would be immediately vaporized. The solution to this problem has been to use the fact that the plasma particles are charged; because they are charged, they can be trapped using magnetic fields instead of material containers. In other words, the plasma can be confined within a magnetic bottle. This method gave rise to a multitude of difficult technical problems, many of which have now been solved. We've made steady progress toward attaining the combination of temperature, density and duration that will yield a self-sustaining energy-producing reaction. But many years and many billions of dollars have been invested to reach the present state of progress. Although we have learned a lot about the physics of plasmas and have achieved many milestones toward practical fusion energy production, we are at best several decades and many more billions of dollars away from this elusive goal. But the effort continues because the fuel for this energy is cheap, relatively non-polluting, and virtually limitless.

Claims of Fusion at Room Temperature

Such was the background for the dramatic announcement (in 1989) that two electrochemists (M. Fleischmann and B. S. Pons) had achieved nuclear fusion at room temperature, using equipment no more complicated than a standard electrochemical cell (which is little more than two pieces of metal, called electrodes, sitting in some liquid, called electrolyte, and connected to a battery). Fusion occurring at room temperature in an electrochemical cell was quickly dubbed "cold fusion." One of the electrodes was made of the metal palladium and the electrolyte contained some heavy water, that is, the hydrogen isotope deuterium was present. Palladium metal is well known to be capable of absorbing large amounts of hydrogen (or in this case, deuterium), the way a sponge absorbs water. The idea of cold fusion is that, under the conditions of the electrochemical reaction taking place in the cell, the amount of deuterium in the palladium electrode becomes so great that the nuclei become close enough to undergo fusion. It's difficult to understand how or why this should happen. But, according to proponents, experimental evidence indicated that cold fusion was in fact happening. In particular, large amounts of heat were measured, which the chemical reactions in the cell couldn't account for.

The results were announced at a press conference before being published in any scientific journal, and the media gave the story a great deal of coverage. Cold fusion was presented as a solution to all of our energy problems (which a cheap and simple way to achieve fusion would indeed be). The experiments were also presented as a revolutionary scientific breakthrough, since the new results seemed to contradict a great deal of well-established science. There was also a human-interest side to the story: the triumph of lonely innovators working on a shoestring budget, where teams of experts with vast resources had failed. What the media did not emphasize was how little documented evidence for cold fusion really existed. Following the original announcement, several research groups quickly started working to test the cold fusion claims. The situation was initially very chaotic and confused. Different groups made conflicting claims of positive and negative results, sometimes followed by retractions. This confusion was probably the result of too much haste and carelessness, brought on by the excitement of the extraordinary claims and their societal implications. As time went on and the results of more careful and well-controlled experiments became available, few research groups found any evidence for cold fusion in the end. The few experiments that did find such evidence were inconsistent and not reproducible. Let's examine this scientific evidence in more detail.

Experimental Evidence

What sort of measurements would indicate that nuclear fusion was taking place? One measurement, which was widely emphasized, is excess heat in the electrochemical cell, indicating that energy is being produced. In addition, the fusion process results in a number of measurable radiations (gamma rays and neutrons) and substances (helium isotopes and tritium). Measurement of these fusion products would indicate that fusion is taking place. The amounts of these fusion products, relative to each other and to the amount of heat produced, can be predicted. This is a crucial point, because the experiments should not only find the fusion products but should find them in the *correct amounts*. Let's take a closer look at this issue, because much of the argument over the validity of cold fusion turns on this point. Figure 7 is a schematic illustration of the three ways in which deuterium can undergo fusion. Protons are indicated by plus signs (+) because they have a positive charge, and neutrons by empty circles because they have no charge. Deuterium is represented by a proton and neutron together, the helium nucleus by two protons and two neutrons, the lighter isotope of helium by two protons and one neutron, and tritium (another isotope of hydrogen) by one proton and two neutrons. In all of these reactions, fusion of deuterium occurs in a way that conserves charge. From many years of experiment and from well-understood theory, we know that the first two reactions are about equally probable, while the

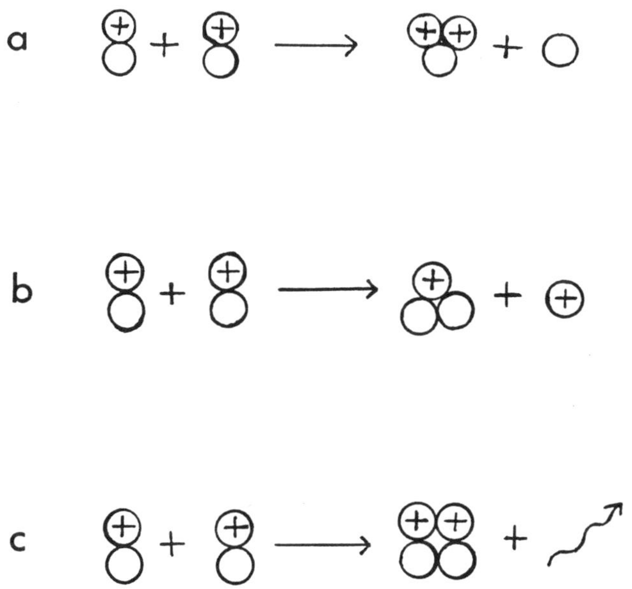

Figure 7. Three possibilities for the nuclear fusion of two deuterium nuclei (deuterium, or heavy hydrogen, is an isotope of hydrogen with one additional neutron). (a) Two deuteriums fusing to produce a helium isotope and a neutron. (b) Two deuteriums fusing to produce tritium and a proton (tritium is another hydrogen isotope, with two additional neutrons). (c) Two deuteriums fusing to produce helium and a gamma ray. This reaction occurs very infrequently compared to the other two.

third reaction is about a million times less probable than the other two. We also know how fast these reactions must proceed in order to produce a given amount of energy (the alleged source of heat in cold fusion).

How do the results of cold fusion experiments compare to these reactions? Results were mixed. Some experiments detected no fusion products (neutrons, helium, etc.) at all, while other experiments detected small quantities of one product or another. Critics have questioned whether those small quantities were really fusion products, as opposed to detector artifacts or naturally occurring contaminants (in other words, phony data). Leaving such questions aside, however, the amounts of the fusion products were far too small to account for the heat that was measured. Also, the amounts were not consistent with the probabilities of the various fusion reactions shown in Figure 7. To make a crude analogy, it's as if someone told you about going on a 2000-mile trip, showed you gasoline receipts consistent with driving 2000 miles, but had an odometer reading increase of only 5 miles on the car. In the case of cold fusion, the inconsistency is dramatic. Some of the people doing the experiments would have received massive lethal doses of radiation to account for the heat, instead of the barely detectable amounts of radiation they reported.

Experiments are always difficult to do. Sophisticated equipment can easily produce readings that are wrongly interpreted; this is why scientists criticize each other's work and try to reproduce experiments. In this case, detecting neutrons is difficult for two reasons: There are always some neutrons around anyway (called background), and detectors sometimes say that the number of neutrons has increased when it hasn't (called drift). Even measuring excess heat isn't as easy as it sounds. Temperature differences across the cell, if they are not measured properly, can make the amount of produced heat look bigger than it really is. Experiments that account for such possible spurious results are better experiments. We say that such experiments are well-controlled experiments, or that they have better controls. The experiments in cold fusion that had the best controls detected no fusion products and little or no excess heat.

Evaluation

Claims for the observation of cold fusion were based on a considerable amount of excess heat and a minute amount of fusion products. Two mysteries therefore needed to be explained. The first mystery is how fusion can occur at all under the conditions of the electrochemical cell (i.e., how do the nuclei get close enough together?). The second mystery is why there are not enough fusion products to account for the heat produced. Several theorists attempted to explain these two mysteries, but the proposed explanations all suffered from the same problem: they were all ad hoc explanations. An ad hoc explanation is an explanation that is not based on anything, an explanation where you just make it up as you go along and use any assumptions needed to achieve the result you want. No coherent theory that really explains the results has ever been proposed to account for cold fusion. A highly developed and interconnected set of theories and experiments, on the other hand, has evolved over fifty years to give us a coherent picture in which cold fusion is not possible.

But, you may object, experiment should be the ultimate authority here; theories must change if experiments contradict them. This is true. But consider the quality of the experimental evidence. None of the experiments claiming evidence for cold fusion have been reproducible. The most carefully done experiments have seen no indications of fusion at all. Even experiments done by proponents of cold fusion have results that conflict with each other in their details. In addition, the experiments are known to be difficult to perform properly, and known to produce phony results sometimes looking much like those obtained by the cold fusion proponents. Is this kind of evidence sufficient to spark a major scientific revolution? The answer to this question is, in the end, a judgment (see chapter 14). The judgment of the mainstream scientific community has been clear: cold fusion does not exist. Few scientists are even paying attention to the question anymore, regarding it as a closed issue. A very small number of scientists consider cold fusion to be real, and they are still working in the area. If these scientists could ever make a convincing case, the mainstream judgment would change. Alternatively, the number of cold fusion proponents may dwindle over the years until it reaches zero some day, and the issue will then die.

§2. PARAPSYCHOLOGY

Parapsychology is the study of various extraordinary abilities ascribed to the human mind. Examples of such paranormal abilities are these: acquiring information without using the senses (extrasensory perception, or ESP); causing something to move without any physical mechanism (psychokinesis); and knowing in advance that some unpredictable event will happen (precognition). Many people (scientists and nonscientists alike) vehemently deny the very existence of all these things and believe that parapsychology is bogus and pseudoscientific. The proponents of parapsychology, on the other hand, believe there is good evidence for the existence of such paranormal abilities. We'll take a look at some of this evidence and discuss how it might be interpreted. We'll also reconsider the question of how to judge whether a field is scientific or not: Should our criteria be based on the methods used or on the subject matter studied?

Historical Context

Folk beliefs about paranormal human abilities are found in almost all cultures and are usually rooted in an irrational worldview. These common folk beliefs are a kind of "prehistory" for parapsychology, but they have little to do with our present discussion. The involvement of scientists in paranormal studies began in England, near the end of the nineteenth century. A quasi-religious movement known as Spiritualism had become prominent (in both the United States and England) at that time. The main activity of Spiritualists was (allegedly) communicating with the spirits of the dead, but this activity was usually accompanied by strange sounds and lights, furniture moving on its own, and other weird things. Much of that display, not surprisingly, turned out to be trickery and fraud. But a number of scientists became interested in these seemingly paranormal phenomena, and they began to engage in a scientific study of Spiritualism. The Society for Psychical Research was founded to provide an organized forum for such studies (a similar American Society was founded shortly afterward). Several prominent scientists and scholars participated in this work. A good deal of their effort was devoted to unmasking fraud, but a number of cases were judged by the scientists to involve genuine effects. These cases were investigated, but in the end it was too difficult to exert proper scientific controls. Interest in this kind of study gradually faded over a number of years.

Parapsychology in its modern form began in the 1930s at Duke University, under the direction of J. B. Rhine and L. E. Rhine (even the word "parapsychology" comes from this source). A new research method was developed. Instead of looking at the spectacular claims of prominent Spiritualists, the Rhines looked for evidence of weak paranormal abilities in average people. In a typical experiment, a subject might look at a random series of images on cards, while a second subject (who can't see the cards) tries to identify which card is being viewed. If this second person can identify the correct card image more often than expected based on random chance guessing, the result is interpreted as evidence for paranormal abilities. A major innovation of this method was the central role given to statistical arguments. Considerable work has been done using such methods, initially at the Duke laboratory and later at several other institutions. Early work was criticized for having insufficient controls (e.g., people might identify cards by seeing smudges on the backs or torn corners). In response to this criticism, experimental designs were improved. Statistical methods were also criticized and, as a result, improved. A professional association was founded, and several journals were established to publish the results of these studies. These same research methods (i.e., looking for small paranormal effects using statistical techniques) are still prominent in present-day parapsychology. Several innovations have been added during the last several decades, such as the increasing use of automated data collection (which is intended to avoid bias and fraud).

So far, I have only been relating the history of the practitioners of parapsychology research. Part of the history of parapsychology, however, is the criticism its practitioners receive from people outside the field. During every phase of the historical sketch I've outlined, critics have said: "These alleged paranormal abilities simply don't exist. Parapsychology doesn't have any subject matter that is real. The results that have been published can all be accounted for by either fraud or careless experimental design." Such charges have been made against the work of each generation of parapsychologists throughout the history of the field.

Reasons for the Controversy

I think the major reason that parapsychology is so severely criticized can be stated very simply: The existence of the paranormal abilities that parapsychologists study is contradicted by the many well-established results of the natural sciences. In other words, the mainstream sciences (physics, chemistry, and biology) look for and find lawful regularities and identifiable causal mechanisms in the workings of the world. In the understanding that these sciences have achieved, there is no way for human minds to affect the world (or other minds) directly, without the use of muscles; or for human minds to gain information from the world (or other minds) directly, without the use of the senses. In the words of one critic, science is a "seamless web," and we can't simply graft the paranormal onto it without disrupting the entire structure. Since this seamless web of interrelated ideas and results is so well confirmed (by countless experiments and observations made by many thousands of scientists over hundreds of years), then we may dismiss the results of parapsychology without further consideration.

Why is this reasoning not compelling to the advocates of parapsychology? There are at least three replies that advocates make to such reasoning. First: Science grows and develops by learning new things and incorporating these new things into a broader understanding; this might eventually happen with paranormal phenomena, even if we can't see how it could happen at the moment. After all, revolutionary changes have occurred before in the history of science. Second: The empirical evidence that paranormal phenomena exist is overwhelming. In the face of empirical observations supporting the phenomena, we can't simply deny the existence of the paranormal because it doesn't fit our current understanding. Third: The results of parapsychology may not actually contradict our current scientific understanding, because certain aspects of quantum physics appear to be consistent with it, for example, nonmaterial information transfer. Needless to say, critics of parapsychology are not convinced by any of these replies. A relationship between quantum physics and psi ("psi" is a general term coined to describe paranormal phenomena and the causes thereof) is highly speculative at best. Any role for quantum physics in explaining psi would be even more controversial than the existence of psi (the very thing it's supposed to justify). Regarding a revolution in scientific thinking to accommodate psi, critics point out that such revolutions only happen when there is a pressing need. In other words, revolutions occur when so many anomalies have accumulated that we can no longer use current scientific ideas productively. If the empirical evidence for psi were indeed overwhelming, then we would be forced to undergo a revolutionary change in our thinking. But the evidence for psi is only meager and weak.

Evidence

You can see that the key question now has become this: How strong, really, is the empirical evidence for the existence of psi? The advocates and critics of parapsychology answer this question in exactly opposite ways. We can't even attempt a complete and detailed review of the parapsychology literature. The number of published studies is very large, and each study would require a critical evaluation. For many studies, a secondary literature of criticism and rebuttal already exists, and we'd also have to look at that. Instead, I'll just describe a few of the studies that parapsychologists consider exemplary. My main interest isn't in settling the issue (even if that were possible); my main interest is in showing how to think about scientific evidence.

The early work with guessing-card images provides several illustrative examples. In Rhine's laboratory during the 1930s, a series of tests were run with J.G. Pratt and H. Pearce. There is little question of subtle sensory clues in this case because the two people were in different buildings (using synchronized watches to time the guesses with the card choices). The cards had five different images (a standard technique in such work), so guessing at random would presumably result in correct guesses for about 20% of the attempts. A lot of data was collected (1850 attempts), and Pearce guessed right on about 30% of these attempts. A later card-guessing series from Rhine's laboratory involved a subject named W. Delmore. Using ordinary playing cards (i.e., 52 images instead of 5), the experimenter placed each card in an opaque envelope and showed the envelope to Delmore, who tried to guess which card was in it. Once again the number of attempts was large (2392). In this case, we would expect correct guesses at random in about 2% of the attempts. Delmore guessed correctly about 6% of the time. The probability of the results observed in these two experiments is vanishingly small. However, it's almost impossible to rule out trickery completely in experiments of this type. Because of the problems with card guessing (e.g., the danger of sleight-of-hand being used), more recent work has often employed an electronic random-event generator. This device is often based on the radioactive disintegration of atoms as its source of random events. In practice, the device might have four possible states that it can choose among randomly, and then indicate its state to the subject by turning on lights. The subject tries to predict which state the device will choose (precognition) or to influence this choice (psychokinesis). The subject registers the prediction by pressing a button; the device records the prediction of the subject automatically, and keeps a running total. Random-event generators are able to acquire large amounts of data quickly and easily, and make cheating difficult. In general, the effects that have been observed using these devices have been small; but because the number of trials has been so large, the results have been statistically significant. For example, one subject predicted correctly 27% of the time (instead of the 25% expected by random guessing). This isn't much, but based on the 15,000 attempts that were made, we would expect such a result only once every 100,000 experiments. Critics of work with random-event generators have questioned whether the device really does produce genuinely random choices, since proof of this randomness is not always included in the published work. Also, the subjects sometimes do worse than random guessing instead of better, which is confusing. Some results with random-event generators, however, have now been replicated by independent investigators.

Our final example illustrates a different kind of research method. Subjects are first asked to relax, and note the imagery that enters their minds. Meanwhile, an experimenter in another room is concentrating on some particular image (which might be anything, e.g., an art print). The test is whether the subject's mental imagery is related to the image the experimenter concentrates on, and whether the subject can identify this image when presented with a number of choices. A typical procedure in these experiments is to promote a relaxed state in the subject with diffuse light and white noise. After about 50 such experiments had been performed, results were pooled for the 39 of them in which the same research methodology was used: Four pictures were presented to the subject, and only a correct choice was counted. (In some other studies, choosing a wrong but similar picture was given partial credit.) The subjects would then be expected to choose correctly 25% of the time by chance. The results of these 39 studies, reporting data from 1,193 subjects, were combined. The subjects chose correctly more than 30 percent of the time, which once again is extremely improbable. Some of the criticisms of this work include problems with randomizing the pictures, and problems with subjects receiving sensory information (e.g., if the same picture that the experimenter handled is used as one of the subject's choices). Also, only about half of the individual studies obtained positive results, making replicability an issue.

Evaluation

There are two questions to consider, and I think the two questions should be considered separately. Is parapsychology a science? Are the paranormal effects studied by parapsychology real effects? I first need to justify my opinion that these are really separate questions, because even this opinion is itself controversial. My reason for this belief is that science is a way of investigating and understanding a phenomenon, not a specific body of knowledge concerning that phenomenon. This subtle point is easy to overlook because in a mature science (like optics or botany) our level of understanding has become so high that the methods and subject matter of the science merge together; the fact that light has different wavelengths and the act of making a rainbow with a prism become inseparable in our minds. The early stages of a science are quite different. In the early history of chemistry, for example, we once thought that a burning material was losing a substance called "phlogiston." We now know that burning materials are combining with oxygen, and that phlogiston does not exist. Many chemists (e.g., the highly respected Joseph Priestley) did experiments to study the properties of phlogiston. No one has suggested that these experiments were not science because phlogiston does not exist. So, it is possible for a science to study effects (at least for a while) that are not real. Parapsychology might still be a science, regardless of whether paranormal effects exist or not. Of course, I haven't answered our question yet, I've only tried to convince you that I can ask the question. Is parapsychology a science?

Although several eminent authorities would disagree with my assessment, I believe the answer is yes. My reason for this answer is that the parapsychologists, in the published works which I've read, accept the ground rules of science: Assertions need to be tested by experiment; flaws in experimental design must be corrected; the generally accepted rules of logical inference should be followed; experimental results should be replicable by independent investigators; and, the ultimate goal of the endeavor is a coherent understanding of the observations. Individual parapsychologists have sometimes made unwarranted statements that violate one or more of these ground rules, and those statements can legitimately be used to argue against my position. But isolated examples are not convincing. The proper question is whether the field as a whole (as an institution) condones such unwarranted statements. There is another argument against parapsychology as a science, and this argument merits serious consideration. The understanding achieved within any science must be consistent with all other sciences. This is the "seamless web" argument that I discussed earlier. We can't have a valid science with an inbred worldview, isolated from the scientific community as a whole. To give parapsychologists more time in which to integrate their results into a broader framework may not be entirely unreasonable, however. Critics will answer that parapsychology has had many years to pursue this goal and has come up empty-handed. But perhaps this is an argument that the field is an unsuccessful science, not an argument that the field is not a science.

We now come to the second question: Are the effects real? We face here a situation similar to that found for cold fusion. A small number of committed workers in the area are convinced that the evidence demonstrates real effects, while a large number of mainstream scientists think this conclusion is nonsense. We've looked briefly at some of the evidence marshalled by the parapsychologists. How should we evaluate this evidence? The key issue here is the burden of proof required. The more radically an idea diverges from well-established knowledge, the more stridently we demand ironclad proof that the idea is right. Since the existence of psi would be a revolutionary change of unprecedented proportions, the evidence for psi must be subjected to unprecedented critical scrutiny. This evidence hasn't yet stood up to the stringent tests required, despite the fact that more modern experiments have eliminated many of the initial objections to their work (such as the easy availability of ways to cheat). The unusually high demand for proof (which is justified by the extravagance of the claims) accounts for the continued skepticism of the scientific community, even when parapsychologists believe that their evidence should suffice.

§3. COMPARISONS

An important similarity between cold fusion and parapsychology is the problem with replicability, which both fields have. If two scientists can't do the same experiment and achieve the same result, we are bound to have confusion and controversy. After a body of well-documented and well-replicated work starts to emerge, controversies tend to die out. In the case of cold fusion, advocates claimed that the problem with reproducing their results was that the experimental techniques of the critics were not right. But this claim couldn't withstand scrutiny for very long, because cold fusion advocates were obligated to specify the right procedures in detail. In the case of parapsychology, the replication problems are in a different category; these problems may be due to the individual differences among the human subjects. If an alleged psi ability is found in experiments with a particular subject, how can another experimenter test this claim without having the same subject? Although this problem might be unavoidable for individual experiments, aggregate results of collections of experiments should be reproducible.

A major difference between cold fusion and parapsychology has been in their historical developments. Cold fusion burst dramatically upon the scene, caused a frenzy of activity, and then almost died out. Parapsychology, in contrast, has had a small but relatively stable number of investigators for over a hundred years. The reason for the initial burst of cold fusion activity is undoubtedly the shock of the initial announcement combined with the social and economic implications of plentiful energy. The reasons this activity ended are related to the string of failures the field endured. The reasons for continued activity in parapsychology are unclear. Proponents argue that the successes of the field continue to attract new investigators each generation, while critics maintain that the widespread and irrational urge to believe in psi accounts for the continued activity of the field.

A rather superficial similarity is that both fields have been labeled pathological science. This label is a term coined by the famous chemist Irving Langmuir to describe the activities of scientists who study nonexistent phenomena by deluding themselves into observing things that aren't there. A classic case cited by Langmuir is the study of N-rays. Around the beginning of the century, many scientists studied N-rays intensively for several years. N-rays don't exist, except (apparently) in the imaginations of those scientists. Cold fusion and parapsychology do share one characteristic that Langmuir attributed to pathological science, namely, they both study effects that are small and difficult to observe. But neither field meets all of Langmuir's criteria, and it's not clear that using this term accomplishes anything. Another similarity between these two fields is that proponents of both fields call for revolutionary changes in our scientific thinking. We have already discussed the basis, in each field, for demanding drastic changes in our understanding. Likewise, we have discussed the case against drastic changes, both in general terms and also for each field in particular. But there is also an interesting difference between these two fields, namely, in the kind of changes demanded. Cold fusion demands specific changes in specific, highly studied physical systems. Parapsychology, on the other hand, calls for changes that are far-reaching but only vaguely defined.

FOR FURTHER READING

Parapsychology: Science or Magic?, by J. E. Alcock, Pergamon Press, 1981.
Polywater, by F. Franks, MIT Press, 1981.
Foundations of Parapsychology, by H. L. Edge, R. L. Morris, J. H. Rush, and J. Palmer, Routledge and Kegan Paul, 1986.
"On the Nature of Physical Laws," by P. W. Anderson, *Physics Today*, December 1990, p. 9.

"Nuclear Fusion in an Atomic Lattice: An Update on the International Status of Cold Fusion Research," by M. Srinivasan, *Current Science*, April 1991, p. 417.

Parapsychology, by R. S. Broughton, Ballantine Books, 1991.

"A Question of Mind Over Measurement," by R. G. Jahn, *Physics Today*, October 1991, p. 14.

Cold Fusion: The Scientific Fiasco of the Century, by John Huizenga, University of Rochester Press, 1992.

At the Fringes of Science, by Michael W. Friedlander, Westview Press, 1995.

EPIGRAPH REFERENCE: Roger Zelazny, *Lord of Light*, Avon Books, 1969, p. 30.

Chapter 14

VERY ABSTRACT QUESTIONS: THE PHILOSOPHY OF SCIENCE

> "And you," I said with childish impertinence, "never commit errors?"
> "Often," he answered. "But instead of conceiving only one, I imagine many, so I become the slave of none."
> *(from* The Name of the Rose *by Umberto Eco)*

SCIENTISTS go about their work, gathering and interpreting data in order to better understand nature. Rarely does a scientist question the basis for this work. What do we mean by "better understanding"? How do we know our methods are valid and our interpretations correct? Why should we even assume that nature can be understood at all? These are questions of a type that scientists rarely ask. Asking this kind of question is the task of the philosopher of science. Philosophers of science have taken a variety of approaches. For some, the goal has been to ascertain the true nature of things, which the methods of science can only hint at. This metaphysical outlook has come under strong attack by other philosophers, who claim that only direct observations have meaning and that the job of science is to find regularities in these observations. In this view, the role of the philosopher is a sort of gatekeeper, to bar metaphysical entities from science. One school of thought assigns to philosophy the work of determining normative methodological rules for the sciences. In other words, philosophers need to tell scientists how they ought to judge between competing theories; select valid evidence; construct explanations with desirable properties; use inductive and deductive logic; determine the truth or falsehood of a claim; and so on. This view was predominant during the first half of the twentieth century, but has since been challenged (primarily by thinkers influenced by the history of science). Historical studies indicate that scientists rarely behave according to the rules prescribed by the philosophers, so what meaning do such rules really have? In this chapter, we'll take a brief look at these controversies, along with a variety of other interesting issues, acquiring a broad overview of the philosophical landscape as we go.

§1. EMPIRICISM AND RATIONALISM

The word "empirical" means founded upon experience and observation. Empirical observations are certainly a necessary ingredient of science, and one might even say that empirical results are the foundation upon which scientific thinking is built. But can science be built on empirical results alone? Although some people might answer yes to this question, such people are in the minority these days. A brief historical sketch reveals rising and falling fortunes for empiricism in European thought. The Greeks were rather more inclined toward metaphysical speculation than empirical observation; but there was also a strain of empiricism in Greek thought, as in the biological observations of Aristotle and in the attention paid to astronomical observations. Empiricism waned during the Middle Ages, as thinkers relied more on the authority of the ancients, but it became prominent in the sixteenth and seventeenth centuries as part of the new learning advocated by natural philosophers. Galileo, for example, effectively used new observational discoveries (Jupiter's moons and the phases of Venus) to argue for the Copernican system (see chapter 5). In England, Francis Bacon enshrined empirical observation as the basis for all progress. Bacon is a sort of patron saint of empiricism; he envisioned the workings of science as primarily a systematic gathering of empirical facts. These facts would then be sifted and organized so as to arrive at general conclusions. Bacon's ideas remained very influential for many years, but the actual practice of science developed a more theoretical strain in parallel with such empiricism. Currently, most people would surely say that empirical observations and theoretical constructions must mutually reinforce each other in order for science to progress.

There is an important problem with basing scientific conclusions solely on empirical facts, namely, the so-called problem of induction. Induction, or inductive logic, basically means that we can conclude that something will always happen because it has always happened before. ("I know the sun will rise tomorrow, because it has risen every day before until now.") This logic is at the core of any attempt to prove a conclusion by empirical observation, because any such set of observations must be finite in number and yet the conclusion is meant to be general. To put it more grandiosely, we wish to make a universal statement based on an isolated set of observations (see chapter 7 for a more practical discussion of induction). The problem with this reasoning is that there is no way we can guarantee that it is true. The fact that something has always happened before does *not* insure that it will always happen. We'll return to this issue in §4.

At the opposite pole of thought is the doctrine of rationalism. The champion of rationalism was René Descartes, who distrusted empirical information as a secure foundation for knowledge because the senses can be fooled. We all know that we can be seriously misled by optical illusions and so on. If the evidence of our senses can't be trusted, what can? Descartes believed that deductive logic, exemplified by mathematics, was the surest foundation for knowledge. Ultimately, then, science should be based on our minds, not our senses. Rationalism, taken to an extreme, proclaims that truth can be apprehended directly by our minds. We only need to use the proper thought processes, identifying manifestly true first principles, and deducing their consequences; no empirical input is needed in this view. Few people still accept this extreme view. Historically, neither empiricism nor rationalism has proven adequate by itself as a philosophical foundation for science. Perhaps the most important result of Descartes' program was to highlight the value of mathematics in the sciences.

Logical Positivism

The marriage of empiricism with mathematical logic was completed in Vienna around 1930 by a group of philosophers who had developed a viewpoint called logical positivism. A major tenet of positivism is that only directly observable objects and events should be considered valid scientific subject matter, making it a highly empiricist philosophy. Anything that can't be observed is, for the positivist, unwanted metaphysical baggage. The task of science, in this view, is to ascertain the logical relationships between all of these observables. The network of logical connections built up in this way is a scientific theory, and such a theory is the only "positive" knowledge that we can possibly have. The clarity and precision of positivism were attractive to many philosophers. Nothing is more clear and unambiguous than mathematical logic; and empirical facts are certainly a firm foundation upon which to build science. But there is a problem: many things in the sciences are *not* directly observable. What do we do about these things? (The grandfather of positivism, Ernst Mach, had stoutly denied the existence of atoms because they could not be observed.) To address this problem, the logical positivists demanded "rules of correspondence" between theoretical entities and empirical observations. As long as a concept could be connected to observations by a set of rigid rules, the concept was legitimate. If such rules can't be found, get rid of the concept.

A similar kind of thinking is found in the movement known as operationalism. The key idea here is the "operational definition" of a scientific concept, which is a definition purely in terms of operations resulting in a measurement. For example, Newton had defined time with a verbal statement that he regarded as self-evident, but that was operationally meaningless. Einstein's redefinition of time in terms of the operations with clocks that are needed to measure time turned out to be an important advance. Using examples like this, P. W. Bridgman (an experimental physicist with philosophical interests) formulated the idea of an operational definition. Bridgman claimed that any concepts that couldn't be defined operationally should be considered scientifically meaningless. The spirit of operationalism is quite consistent with logical positivism, and both were heavily influenced by Mach. In terms of the positivist program, operational definitions are a particular kind of correspondence rule. Logical positivism dominated the philosophy of science for decades, and still exerts some influence. But it came under attack around the middle of the century, and has waned in authority since then. We'll see some of the problems with positivism as we go on with the chapter. A different philosophical vision of science, however, had been articulated even before positivism was invented. The brilliant French physicist and philosopher, Henri Poincaré, proposed that science needs to employ "conventions" that are creations of the human mind, not directly tied to empirical facts. These conventions are not merely free creations, however, because they are constrained by our observations of the world in a variety of indirect ways, and sometimes must change with the course of scientific progress. Poincaré, writing near the turn of the century, devised a philosophy of science that in many respects foreshadowed much later developments.

§2. FOUNDATIONAL QUESTIONS

Causality and Determinism

The idea of causality is very important in the sciences but has been notoriously difficult to make philosophically precise. The root word of causality is "cause." If one event or action is the cause of another, they are connected by a causality relation. This sounds simple enough, but what do we mean when we say that one thing causes another (see chapter 7)? For example, what caused my coffee cup to break? I pushed my coffee cup off the shelf. So, my pushing it caused the cup to break. But we could just as well say that gravity pulling it down caused the cup to break; or, we could say that the brittleness of the clay caused the cup to break. This example illustrates two problems: the problem of multiple partial causes, and the problem of logical necessity (i.e., if the cause occurs, the effect *must* follow). In the physical sciences, discussions of causality are restricted to cases where a single cause can be identified that always gives rise to a well-defined effect. In the life sciences and the social sciences, where such a restriction is seldom possible, the concept of causality is still often employed (e.g., a virus causes the common cold, but exposure to the virus doesn't always necessarily result in a cold). Although descriptions based on causality are quite widespread and important in these fields, philosophical analysis of such usage has not been thoroughly worked out. The restriction to cases of single causes and necessary effects is very limiting, of course, but does permit a high degree of logical rigor in the discussion.

Empiricist philosophers, however, have used this criterion of universal conjunction to deny the meaningfulness of causality. As David Hume pointed out, the most that we can ever really know is that one event has always followed another; to then claim that the first event causes the second is merely a bad verbal habit. Night always follows day, for example, but we don't claim that the day causes the night. This isn't the last word, though, because empiricism isn't the only game in town; other philosophers have elevated causality to the status of a metaphysical principle. Philosophers who were heavily influenced by physics adopted a different line of thought. Since you can write down equations and solve them in order to predict the second event based on the occurrence of the first event, then predictability becomes the key issue (see, however, chapter 17). Another word for such predictability is determinism; the first event determines the occurrence of the second. Put differently, the important property of the world is that it's deterministic, not that it's causal. Causality, in this view, is just anthropomorphic language assigning motives to a deterministic sequence.

Chance and Probability

But if the world is deterministic, there should be no random events; and yet, we know that dice rolls and lightning strikes are unpredictable. Even without using the new insights gained in the last few decades (chapter 17), this paradox can be resolved. The prediction of an event requires a complete specification of the state that deterministically gives rise to this event. In practice, meeting this requirement is rarely possible. From this standpoint, the need to use probability in the sciences stems from what we don't (and perhaps can't) know. How can we figure out what the probability of an event is? There are well-defined mathematical rules for combining probabilities. For example, I can calculate the odds in a coin toss of getting five heads in row, if I know the odds of getting a head on each toss. But how do I know the odds of each individual toss coming up heads? How do we figure out the single-event probabilities that we need in order to make combinations? There are two main schools of thought on this question: one subscribes to the principle of indifference, while the other adopts the frequency interpretation.

The principle of indifference is simply stated: if there is no reason to believe that one event is more or less likely than another, then the events have equal probabilities. The odds of getting heads in a coin toss are 50 percent; the probability of any particular face coming up on a tossed die is 1/6; there is a 1 out of 52 chance of drawing any particular card from a deck; and the odds of finding a particular air molecule in one volume of a room are the same as finding it in any equal-sized volume. The adherents of the frequency interpretation, being empiricists, find this idea too metaphysical. They instead define the probability of an event as the ratio of the number of times this event occurs to the number of possible times it could occur. In other words, we toss many coins, and count the number of heads that turn up. The disadvantage of this technique is that it's often impractical to carry out, but the obvious advantage is that you find out if your presuppositions are wrong (for example, you may be using loaded dice).

The arguments between these schools of thought tend to be highly technical. The questions are important, however, because many sciences are based on chance and probability. Prominent among these are statistical mechanics, genetics, the theory of experimental errors, quantum physics, and much of the social sciences. In addition, some strains of modern philosophy consider scientific theory validation to be primarily a matter of increasing the probability of correctness, since we have no method of assuring certainty. Finally, questions of determinism and chance in the sciences influence the broader philosophical discussions of free will and destiny in human affairs.

Space and Time

We experience our lives in space and time. Long before the dawn of a scientific worldview, philosophers (not to mention poets and mystics) were deeply engaged by questions about the nature of space and time. Since all natural processes occur in space and time, these concepts are fundamental to science. (Examples might include the formation and erosion of mountain ranges, or the geographic distribution of plants and animals during a period of climate change.) But the use we make of space and time in many sciences is pretty straightforward, with little need for philosophical analysis. An exception to this statement is physics, in which space and time become themselves objects of study. Newton's descriptions of space and time as absolute properties of reality, unaffected by either matter or consciousness, were in fact metaphysical presuppositions. The writings of Kant put this metaphysical view of space and time on a more rigorous philosophical foundation. Put simply, the properties of space are given by the geometry of Euclid, and time is a separate property that exists independently of space. Einstein discovered that these metaphysical properties of space and time are incompatible with the logic of modern physical theory (electrodynamics and mechanics don't fit together properly), as well as certain experimental observations. In response, Einstein reformulated our conceptions of space and time. In Einstein's relativistic worldview, space and time are welded together into a single four-dimensional spacetime, whose non-Euclidean properties depend on the matter that resides there. One challenge of twentieth-century philosophy of science has been to find a good metaphysical foundation for this revolutionary conception (Ernst Cassirer's work here is noteworthy).

A second major question concerns the directionality of time, the so-called arrow of time. Past, present, and future have always been a mystery worthy of consideration by philosophers. In science, part of the mystery stems from the role that time plays in the equations of physics. In the physics of both Newton and Einstein, these equations don't depend on the plus or minus sign of the time; the same motions are predicted either way, only the direction of the sequence of events will be reversed. In other words, the universe could just as well be running either forward or backward, as far as the equations are concerned. Of course, we always experience the universe as running forward, from past to future. Why should this be? Some insight into the question comes from statistical physics (see chapter 17), but philosophers are still pondering these deep issues.

Reductionism and Emergence

To illustrate what reductionism means, let's consider an example. Kepler's laws tell us descriptively how the planets move around the sun (see chapter 5). Newtonian dynamics tells us, in theory, how any object should move. If we apply Newtonian dynamics to the case of the planets and the sun, we find that Kepler's laws result from the analysis; we are able to explain, in broader terms, why they work. In other words, we have reduced Kepler's laws to a special case of dynamics. Different degrees and levels of reduction can be found in the sciences. In our example, there are two levels of reduction; first, Kepler reduced reams of observational data to a small number of empirical rules, and then Newton reduced these empirical rules to a special case of a general theory. This sort of reduction occurs all the time in the sciences; in a sense, we're really just carrying out a principle aim of science, subsuming the particular to the general. But under what conditions can such a reduction be accomplished? Can any scientific result be reduced in this fashion? If so, isn't all of science then ultimately reducible to some final irreducible theory? These are some of the philosophical questions concerning reductionism. Whether any science can be reduced to some other (more fundamental) science is a hotly-debated question.

Let's first consider a classic example of one science being reduced to another. The science of thermodynamics is entirely based on three general laws, which were originally distilled purely from observations and experiments on heat and bulk materials. In other words, thermodynamics says nothing about the microscopic properties of materials (or heat). The science of statistical mechanics, on the other hand, assumes that materials are made of atoms obeying the laws of dynamics. Since we cannot in practice predict the individual motions of the vast numbers of atoms, we use probability theory along with dynamics to make predictions for averaged values. Remarkably, theorists discovered that these averaged values correspond to the variables of thermodynamics. We can now understand thermodynamic results based on a microscopic picture, and so we can say that thermodynamics has been reduced to statistical mechanics. This accomplishment is one of the great victories of reductionism.

But the atoms themselves are made of subatomic particles, which are studied by high-energy physics. Does this mean that thermodynamics can be reduced to high-energy physics? Let's expand this line of thought: Chemistry studies reactions between atoms and molecules, forming and breaking chemical bonds. These bonds are formed by electrons, subject to the equations of quantum physics. So chemistry is reducible to physics. Biology, however, studies organisms that are made of chemicals. Psychology studies human thinking, which occurs in the brain, a biological organ. Sociology studies societies, which are made of humans. We can then make the claim that sociology is reducible to physics through a kind of hierarchical chain of reductions. There are, however, a number of problems with this absurd claim. An important technical problem is that successful reductions must involve clear and precise correspondences between all terms in the two theories or sciences. If this condition is not satisfied (it rarely is), then each science is an independent thought system with its own terms and structure, not necessarily reducible to anything else. On a practical note, the "reducing" science can't predict all of the results found in the "reduced" science in most cases (not even chemistry and physics, where reductionism seems fairly viable). Favorable synthetic reaction pathways are more often found by working in the lab, not by solving quantum physics equations.

Another argument against extreme reductionism has been advanced by the philosophical movement known as holism. The basic idea here is that a whole may be more than the sum of its parts. A complex system may exhibit emergent properties that can't be predicted by analysis of the component parts of the system. But a reductionist program requires that the complex systems of one science can be understood in terms of the component parts in its "reducing" science. For example, a biological cell must be understandable based solely on the chemical reactions occurring in the cell. If the cell has emergent properties not predicted by these chemical reactions, biology can't be reduced to chemistry. Emergence, as a philosophical doctrine, attracted both interest and controversy (reductionists criticized it as disguised vitalism). Recently, however, the idea of emergent properties has received a boost from within science itself, as the study of complex systems has progressed (see chapter 17). In any event, few philosophers (or thoughtful scientists) still subscribe to the radical form of reductionism (although it's still often found in introductory textbooks). The sense in which reductionist programs are still valid (and to what extent) is an active question in the philosophy of science.

§3. Epistemological Roots

How Do We Know?

The branch of philosophy concerned with problems of knowledge is called epistemology. How do we acquire knowledge of the world? What does it mean to know something? How do we know that our knowledge is true? How is our knowledge of something related to the thing itself? These are the kinds of questions explored by the philosophy of knowledge, epistemology. Epistemological questions are broader than the questions found in the philosophy of science, but are in a sense foundational. The roots of the philosophy of science are deeply embedded in epistemology. Here, we'll take a brief look at some of the epistemological issues that are most relevant to the philosophy of science.

Percepts and Constructs

We often think of perceptual observations as simple facts, not reducible to anything simpler. "I see a tree." We call this an observation, not a conceptual statement. But what I actually perceive, before my mind imposes order on it, is some spatial distribution of colors and intensities. And even these result from various reflected light rays striking the cells of my retina, sending neuronal signals to the brain. The actual perception results, in some way we don't understand, from the complex pattern of interconnected neuronal activity in the brain. It's not simply a matter of "a tree is there" and "I see it." So even in the simplest cases, our perception of the world entails our mind imposing some order on a mass of information produced by signals originating in the world. Put differently: we create constructs based on our percepts. The point is that this is necessary, not optional. We can't avoid the use of constructs in our empirical observations. This doesn't mean that our perceptions are arbitrary, because there are connections between the percepts and constructs, rules which govern how the constructs are created from the percepts. If these rules are well defined, then empirical observation is still meaningful.

In scientific discourse, the terms under discussion (i.e., constructs) become more complex and abstract. Instead of a tree (which I can see), we might be studying photosynthesis (which I can't see). But the process of photosynthesis can be related to the absorption of light and the production of glucose, which I can measure and analyze chemically. My actual empirical observation might only be the pointer reading of the instrumentation I use to measure the absorption or analyze the glucose. But there are well-defined rules (rules of correspondence, in positivist language) that relate the pointer readings to the quantities, and there are clear relationships between the measured quantities and the photosynthesis process. Our facts and our empirical observations are embedded in a dense matrix of interpretation.

Notice that we have now introduced a new element into the discussion. We are now talking about relationships between concepts only (e.g., "glucose is produced by photosynthesis"). The rules of correspondence relate the concepts to observations, but there are also purely logical relationships between the concepts themselves, none of which are directly observable. Constructs can be related to each other by logical relationships, and related to percepts by rules of correspondence. Are we free in science to create any constructs we please? The answer is no. Higher-level constructs may not be unique, but neither are they arbitrary. In order to be useful in science, constructs must have certain properties. We've already seen that constructs must be connected to other constructs by logical relationships; an isolated construct cannot be valid in science, no matter how interesting it may otherwise be. Likewise, we've seen that scientific constructs must ultimately have some connection to perceptions, at least an indirect connection. (Hierarchies of angels and demons with many intricate relationships to each other, but nothing to tie them to perception, are not objects of scientific study.) Constructs should also have some degree of stability; we can't just keep making it up as we go along. They should also be extensible, that is, capable of being generalized. Finally, scientists have a preference for constructs that are simple rather than being unnecessarily cluttered with complications.

Theory-laden Facts

The foregoing considerations bring us face-to-face with one of the problems of logical positivism (and empiricism in general). Empirical facts are already constructs, not "raw" percepts. The terms used in the language of a scientific theory are also constructs (more abstract than the fact-constructs). Now, the rules by which fact-constructs are created from percepts may be (and almost surely are) influenced by the theory-constructs to which they are related. Our empirical facts are not then totally independent of the theories we use to interpret them. As a trivial example to demonstrate the point, suppose I say that empirical observation proves that the sun rises in the east every day. "East" and "sun" are both complicated collections of concepts, involving geographical knowledge, astronomical knowledge, and so on. Surely that statement must mean something very different to us than it meant to the ancient Celtic Druids. How can we then consider it a simple empirical fact? This intertwining of theory and observation is one of the problems with the view that scientific results can be proved by purely empirical means. If empirical observations already have a conceptual component inherently built into them, it is circular reasoning to say that our observations have proved our concepts correct. This situation has been noted by philosophers, who use the term "theory-laden" (coined by N. R. Hansen) to describe the interpretive matrix that accompanies seemingly simple facts. If empirical facts are theory-laden, then the positivist program is clearly derailed. What's not so clear is the extent to which the theory-laden quality of facts impairs our ability to validate a scientific theory by empirical tests. Some thinkers have concluded that objective science isn't even possible under the circumstances, while others believe that only slight adjustments to empiricism are needed. This issue, along with other issues of theory validation, is considered next.

§4. Validation of Theories

Failure of Induction

Logical positivism has a problem, even if you accept the pristine theory-independence of facts: the problem of induction. On the one hand, there is a purely logical problem with inductivism. Any logical proof of the inductivist premise must be itself a proof by induction ("induction has always worked, therefore it is valid"). This obviously is no proof at all, since it assumes the truth of the premise it sets out to prove. On the other hand, there is a more practical problem: the conclusions based on induction do, in fact, sometimes turn out to be wrong. For example, Lavoisier had shown (by the late eighteenth century) that a number of substances are elements, that is, substances that cannot be further broken down or changed to another substance. By the beginning of the twentieth century, this property of elements (unchangeability) had been shown empirically to be true innumerable times. That elements are immutable was taken to be a well-known fact; note that the conclusion is clearly based on induction. Around 1900, radioactivity was discovered, and Rutherford demonstrated that the radioactive decay process is actually a change from one element to another. The previous conclusion, based on induction, had simply been wrong.

Falsificationism

To avoid the problem of induction, we can turn the question around: Once a theory has been proposed, we demand that it make statements that can be falsified by comparison with observation. We then make the observations. If they don't agree, that is, if the statement is false, we pronounce the theory wrong. If the observation agrees with the prediction (statement is true), then we say the theory is confirmed by the observation. But we don't necessarily say that the theory is true. We've made a subtle but important change in our philosophical view concerning the goals of science. In this picture, we accept theories on a provisional basis, as long as they continue to be confirmed. We can never prove a theory is right, only that it's wrong (if a statement is false). This idea has been implicit in science for some time, but as a formal doctrine it is closely associated with the philosophy of Karl Popper.

The criterion of falsifiability serves a number of worthwhile purposes. Primarily, it allows us, when deciding questions of validity, to weed out statements that have no observable consequences. An extreme example is the claim of an Aristotelian opponent of Galileo that he made after he (the opponent) observed through a telescope mountains and craters on the moon. To make this observation consistent with his belief that the moon is a perfect sphere, he hypothesized the existence of an invisible substance that covered the mountains and filled the craters, making a perfect spherical surface. This substance could not in principle be observed, so his claim could not be falsified. Such statements are of little use in the discourse of science.

But the rigorous logic of falsificationism is rarely enforced in practice; if it was, many productive and valid ideas would have died an early death. A famous example is the so-called stellar parallax predicted by the Copernican theory. Stellar parallax simply means that the stars should appear to be in slightly different places as the earth itself moves around from place to place. This sensible conclusion is hard to escape, but the effect was not observed until about 1840, roughly 300 years after the work of Copernicus. Strictly speaking, the theory was falsified. But it wasn't given up, nor should it have been. (The reason parallax was not seen is that the stars are much farther away than anyone suspected, and it awaited the technical development of better telescopes to make the observation.) The point here is that a good theory, which has a number of empirical confirmations and is leading progressively to new ideas and applications, will not and should not be given up even if a few minor discrepancies remain. Falsificationism is still a valuable methodological tool. The confirmation of bold new ideas, when put to the test of falsification, drives the progress of science (as in the case of bending starlight predicted by Einstein's general relativity in 1915 and observed during a solar eclipse in 1919). Equally important, the falsification of well-accepted ideas, as in the radioactive transmutation of elements cited above, also drives the progress of science. So, falsificationism serves well as a framework for thinking about scientific questions, but it doesn't provide us with a universal principle for determining validity in science.

The Quine-Duhem Thesis

In our discussion of stellar parallax, we saw that an apparently disconfirming fact could be accounted for by the overall context of the situation (large distances, limited instrumentation, etc.). This episode is indicative of a more general problem, namely the underdetermination of theories by facts. The general idea is this: Any scientific theory is a complex interlocking set of concepts, observations, definitions, presuppositions, experimental results, and connections to other theories. No single fact is going to be crucial to the survival of a theory. We can illustrate this point with a simple hypothetical example. Suppose I have a theory that predicts that carbon turns to diamond at a certain very high pressure and temperature. I try it, and it doesn't happen. My theory appears to be falsified. But, how do I know what the temperature and pressure really are? To measure extreme pressures and temperatures requires special instruments, which themselves operate according to theories. Maybe *those* theories are false under our conditions, and my diamond theory is correct after all. Empirical facts are not always unambiguous. So, the complexity of many interdependent and theory-dependent terms makes it difficult to frame an unambiguously falsifiable statement. Pierre Duhem argued as early as 1906 that our experiments only test the total structure of a science, not individual hypotheses. W. V. O. Quine later extended and strengthened this idea considerably, resulting in the so-called Quine-Duhem thesis: It's impossible, on purely logical grounds, to falsify a theory experimentally, because any observation can be accommodated by making suitable adjustments.

The Quine-Duhem thesis has generated a lot of controversy. Some thinkers believe the claim is a triviality, while others believe that it utterly destroys the objectivity of science. An important point to keep in mind is that the adjustments we must make to save a theory, while they may be logically possible, may also be unacceptable on other legitimate grounds; the Quine-Duhem thesis does *not* mean that all theories are equally good. The question then becomes this: On what grounds are such judgments made?

Kuhn and Scientific Revolutions

This last question was examined, along with a host of other issues, in a very influential book called *The Structure of Scientific Revolutions* by Thomas Kuhn. Kuhn's work, published in 1962, was based on historical studies of how science is actually done, as opposed to the normative work typical of the philosophy of science before that time. Since the book was published, many critical commentaries on it have been written. Kuhn also continued to refine his own philosophy of science, and a legion of scholars have rather freely interpreted his work to support a wide variety of positions and viewpoints. Very briefly, Kuhn's view is this: Scientific communities operate by sharing a set of assumptions (some tacit and some explicated), techniques, and methodologies, along with a common terminology and worldview. All of this he collectively refers to as a paradigm. Work within a paradigm is called normal science, and consists of filling in details, solving puzzles, and so on. When the number of unsolvable puzzles, loose ends, empirical facts that don't fit in, and so on becomes intolerable, a scientific revolution occurs; after this, an entirely new paradigm is adopted. During periods of normal science, determining validity is relatively straightforward because everyone agrees to the same rules and talks the same language. During a scientific revolution, however, when two different paradigms are competing, it becomes extremely difficult to determine validity based on simple empirical comparisons because different scientists might be talking a different language and even seeing a different world. For example, if I live in a world where combustion occurs by the driving off of a substance called phlogiston, and you live in a world where combustion occurs by the combining with a substance called oxygen, how can we compare the results of our experiments to arrive at a common conclusion? Eventually, the new paradigm becomes established. This new paradigm is determined to be better than the old paradigm by a consensus of the scientific community. The process driving the paradigm shift includes a combination of empirical criteria (e.g., there are fewer falsified statements) and nonempirical criteria (e.g., the new paradigm leads to a period of vigorous activity and progress).

Kuhn's picture is controversial as a general description of scientific progress, but there are certainly some elements of truth in it (at least for some historical eras). Our main interest here is the key role played by consensus judgments of scientific communities. Validity is not established merely by a rational comparison of theories, applying a set of predetermined standards that automatically lead to a unanimous choice. Instead, individuals working in communities try to sort out a maze of theory-laden observations (of varying accuracy and relevance), out of which they formulate a coherent worldview where observations make sense to them. Eventually, the individual judgments coalesce into a consensus judgment of the whole community as a powerfully coherent new formulation emerges. Those who are not convinced become marginal to the process (cranks). Kuhn's work has been attacked because it has been taken to imply that there are no standards, that validity in science is determined by majority rule (the work has also been applauded for the same reason; Kuhn, incidentally, didn't agree with this interpretation). The accusation is that nonscientific criteria, such as political ideology or religious conviction, may play a role in the process. Kuhn's work certainly implies that there is some role for the judgment of scientific communities. But the precise nature of that judgment, and the extent to which it sanctions a kind of irrational antirealism, are very controversial matters indeed (see chapter 15).

Criteria for Theory Selection

A theory that explains more is better than a theory that explains less. If we need to make up a separate explanation for each individual phenomenon, we have not done very well; but if a single theory explains a set of disparate phenomena all at once, then we've accomplished something. The greatness of Newton's mechanics is that it explains the orbits of the planets around the sun, the spinning of an ice skater, and the workings of a grandfather clock, all with the same laws of motion. So explanatory power is one of our criteria for what makes a theory good. What else makes a theory good? Simplicity: a simple and elegant theory, with few assumptions, will be chosen over a cumbersome and ugly theory, even if both explain the observed data equally well. Fertility: a good theory leads to new ideas, new applications, new connections to existing theories, and new refinements of itself (eventually, to even newer theories). A theory that is an intellectual dead end is a poor theory, even if it explains all the current data (vitalism, in biology, might be an example). Finally, a good theory should lead to the prediction of unforeseen results, to something that was not known when the theory was devised. For example, the highly mathematical theory of electricity and magnetism derived by James Clerk Maxwell predicted the existence of radio waves, which no one had ever observed or even suspected. None of these criteria can be justified from an empiricist viewpoint because in each case the "good" theories and the "bad" theories account for all known data equally well. And yet, science makes extensive use of these criteria in theory selection judgments. We might justifiably label such criteria as metaphysical statements, although the word "metaphysical" carries a lot of undesirable excess baggage.

Judgment in Science

Some element of human judgment must surely be a part of the critical process in science (after all, science is done by humans). But the judgments of individual scientists, and the scientific consensus that they formulate, are severely constrained by nature. Scientific theories are not free constructions of the human mind, any more than they are inductive generalizations of simple facts. A rigorous methodology of science that excludes human judgment may or may not be impossible, but it certainly hasn't been accomplished. Yet judgments are made within the context of agreed-upon methodological standards that allow us to employ nature as a reliable (if not infallible) guide. So we have arrived at the following answer to our question of how validity is determined: Out of our initial set of empirical observations and experiences, shaped by our unformulated preconceptions, we begin to develop initial scientific concepts. These concepts are used to organize a growing stock of observations, which are combined with various nonempirical criteria (such as simplicity, fertility, and explanatory power) in order to formulate more refined theories. This last step requires the exercise of scientific judgment on the part of individuals and communities. The refined theories lead to a set of predictions, statements, and claims that should have falsifiable consequences. These predictions, statements, and claims are then tested against empirical observation, with scientific judgment again exercised to resolve any ambiguities that present themselves. The collective result of this judgment is a consensus of the scientific community as to whether the theory is valid or not. This consensus judgment is retained provisionally, since new observations are being made and new theories are being formulated that may challenge it.

§5. SCIENTIFIC EXPLANATION

What is the point of a scientific theory? We've looked in some detail at the question of whether a theory is correct, but what does a correct theory do for us? A theory is intended to explain the phenomena we observe. The question then becomes this: What do we mean by a scientific explanation? Philosophers have had a very difficult time answering this thorny question, and we'll just take a brief look at it. The positivist answer was that an explanation is a deduction from a law of nature. In other words, if I want to explain why some event happens, I start with some totally general premise (a law of nature) that is always true; I then use deductive logic to demonstrate that this event must follow from this premise. For a science that is already in a highly mathematical form (such as physics) this works reasonably well, but it doesn't work so well for less axiomatized sciences (such as geology). Another problem is that it's possible to construct explanations that formally satisfy these conditions but don't actually explain anything. Finally, the idea of a law of nature has never been defined sharply. Yet the basic point of this concept of scientific explanations seems to be on the right track. The essential idea is that an explanation should account for many disparate phenomena by some small number of fundamental premises. This point is still valid even in the absence of all the formal deductive logic conditions, and it remains valid for virtually any science.

Another aspect of scientific explanation is that there are often successively deeper levels of explanation, each level explaining what is left unexplained by the previous level. For example: The properties of a particular element are explained by the position of the element in the periodic table, which brings an ordered regularity to such properties (see chapter 2). The organization of the periodic table itself is explained by the electron shell structures of the elemental atoms. These shell structures, in turn, are explained by the Pauli exclusion principle of quantum mechanics. And the exclusion principle is explained by the symmetries of the quantum mechanical wavefunction (see chapter 18). The successive layers of explanation unfold like an onion being peeled. At the core of the onion is a fundamental principle that cannot be further explained (in the previous example, a symmetry principle). This too is characteristic of scientific explanation; an irreducible level is reached, beyond which no further explanation is possible at that point in a science's development. Future progress might or might not unfold at a deeper level, but until then, the premises of this irreducible level are just taken to be the starting point and accepted as such.

FOR FURTHER READING

The Foundations of Science, by H. Poincaré, Science Press, 1913.

Substance and Function, by Ernst Cassirer, Open Court, 1923 (Dover reprint, 1953).

The Logic of Modern Physics, by P. W. Bridgman, Macmillan, 1927.

Philosophy of Mathematics and Natural Science, by Hermann Weyl, Princeton University Press, 1949.

The Nature of Physical Reality, by Henry Margenau, McGraw-Hill, 1950.

Scientific Explanation, by R. B. Braithwaite, Harper & Brothers, 1960.

The Structure of Science, by Ernst Nagel, Harcourt, Brace & World, 1961.

The Language of Nature, by David Hawkins, W. H. Freeman, 1964.

Philosophy of Science: An Introduction, by Paul R. Durbin, O.P., McGraw-Hill, 1968.

The Structure of Scientific Revolutions, by Thomas Kuhn, University of Chicago Press, 1970.

Can Theories Be Refuted?, edited by Sandra G. Harding, D. Reidel, 1976.

Philosophy and Science, edited by F. W. Mosedale, Prentice-Hall, 1979.

What Is This Thing Called Science?, by A. F. Chalmers, University of Queensland Press, 1982.

Introduction to the Philosophy of Science, by Robert Klee, Oxford University Press, 1997.

EPIGRAPH REFERENCE: Umberto Eco, *The Name of the Rose*, Harcourt Brace Jovanovich, 1983, p. 133.

Chapter 15

QUESTIONS OF LEGITIMACY: THE POSTMODERN
CRITIQUE OF SCIENCE

> Yet in the choice of these man-made formulas we can not be
> capricious with impunity any more than we can be capricious
> on the common-sense practical level. We must find a theory
> that will *work*; and that means something extremely difficult;
> for our theory must mediate between all previous truths and
> certain new experiences. It must derange common sense and
> previous belief as little as possible, and it must lead to some
> sensible terminus or other that can be verified exactly. To
> "work" means both these things; and the squeeze is so tight
> that there is little loose play for any hypothesis. Our theories
> are wedged and controlled as nothing else is. Yet sometimes
> alternative theoretic formulas are equally compatible with
> all the truths we know, and then we choose between
> them for subjective reasons.
>
> *(William James)*

A BATTLE is raging in academia over the issue of whether objective knowledge is possible. The opposing camps are not really very well defined, but in broad terms we might say this: the traditionalists, on one side, favor western culture, values that are absolute, and truth; the postmodernists, on the other side, favor multiculturalism, relativism, and a worldview in which truth doesn't exist. While these caricatures oversimplify the interesting range of issues involved, the ramifications for fields like history, philosophy, and sociology are clear. The sciences have also been drawn into this conflict. Scientists have generally thought of their discipline as objective, free of values, and leading to definite knowledge (as opposed to mere opinion). But the postmodern critique denies the possibility of objectivity and of definite knowledge. Postmodernists point out that science is a product of western culture; if we deny the primacy of western culture, then science is just another claimant (among many) for the validity of its results. An entire area of academia (science studies) has been built up around this basic idea. A number of scientists have been rather vocal in taking exception to these postmodern claims. A point these scientists often press is that many of the postmodern critics of science have little understanding of either the results or the methods of science; for this reason, such critics are poorly qualified to make the claims they do. This attitude came to a head in 1996, when physicist Alan Sokal wrote a parody imitating typical postmodern papers and then published it in one of their leading journals. Leaving aside questions of whether this action was ethical and what (if anything) we might conclude from the episode, it certainly brought the controversy to everyone's attention.

Several murky issues are tangled together in this controversy. Is science merely the tool of an iniquitous imperialistic system? If so, is this situation necessary, or is it contingent? Does the validity of a scientific result depend for its warrant on the justice of the sociopolitical system in which the science is done? Is the achievement of scientific consensus a rational process, or a political process? Are scientific theories accurate reflections of a pre-existing reality, or merely social constructions based on negotiations between interest groups? Are the knowledge claims of science more valid than knowledge produced by other ways of knowing, or are all such claims equally valid? Let's explore these questions and try to untangle some of the relevant issues. One source of confusion stems from mixing together two very different issues: the question of whether science is good or bad, and the question of whether science delivers a correct understanding of nature. The validity of an idea doesn't depend on its moral virtue (or lack thereof). The question of whether science is good must surely depend on what our values are, and that discussion belongs in a different place (chapter 11, which includes a serious consideration of these matters). The second question (is science valid?) is our primary interest here.

§1. THE CONSTRUCTION OF SCIENCE

Is the Definition of Scientific Validity a Cultural Artifact?

One of the most abstract, general, and powerful arguments of the science critics is based on the following premise: We and our thinking processes are products of our culture. Even our most fundamental assumptions are culture-bound, rather than being true in any absolute sense. For example, two of the fundamental assumptions of science are that, in formulating our understanding of nature, observational evidence is paramount and logical coherence is necessary. A cultural critic might argue that these assumptions are really just products of European thought during the last four centuries and need not be universally made. This argument is extremely difficult to counter because any point you make can be dismissed. Contrary evidence is not decisive if the primacy of evidence can be denied. Logical flaws can likewise be ignored if logic itself is merely a cultural artifact. Scientists often point to the success of science in making predictions, but the high esteem we bestow on predictive success is, once again, a product of culture. Along these lines, the philosopher Paul Feyerabend has made the comment that voodoo is no better or worse than science as a way of understanding the world. If we don't get to make *any* presuppositions at all, who is to say that Feyerabend is wrong?

But this argument has force only if we restrict our attention to the most extreme alternatives. I am willing to concede that science has no *absolute* superiority to other thought systems in the sense I've just outlined. Science is indeed the product of a particular culture in a specific set of historical circumstances. However, the alternative to "science is absolutely superior" does not have to be "all modes of thinking are equally valid." Between these two extremes lies a continuum on which we can make and justify judgments. That science is intrinsically superior to any other realm of human endeavor is certainly questionable, but that science is a purely arbitrary construct is at least equally questionable.

Are the Results of Science Social Constructions?

Inspired by the work of Thomas Kuhn (chapter 14), sociologists of science began to look more closely at the process by which scientists arrive at a consensus. They reported that the process is not the purely rational sifting of evidence that we sometimes pretend it is. In fact, scientists are humans, and the workings of the scientific community are a process of social negotiation. Matters of prestige, friendship, and power (which should be irrelevant) do enter into these social negotiations. The question is this: Do these social forces warp, or even determine, the results of science? In the early stages of a scientific discussion, when the ideas are still unclear and the evidence is sparse, these "illegitimate" social factors must surely influence the discussion. The postmodern critics who subscribe to the so-called strong program of social constructivism argue that these influences always permeate the discussion and eventually determine the outcome. Many scientists (and like-minded philosophers) maintain that such social influence is limited by the constraints imposed by nature; there *is* a reality, and it will decide the issues in the end, regardless of prestige and power.

An example illustrates this point. When S. Chandrasekhar was a young and unknown scientist, he performed a set of calculations suggesting that massive stars end their lives by gravitational collapse (black holes, as we now call them). The foremost authority in astrophysics at that time, Sir Arthur Eddington, could not accept such an absurd-sounding conclusion. He thought that something (unspecified) must be wrong with the theory, and he said so. Although Chandrasekhar had the better argument, Eddington's conclusion was almost universally accepted in the scientific community. No observational evidence of any sort existed then, the idea was strange as well as new, and Eddington's prestige was enormous. But today we accept the presence of black holes in the universe with few reservations. This change is due to the many new measurements made in astronomy (see chapter 3), working together with an increasingly sophisticated theoretical understanding of the issues. But this process of changing from a world without black holes into a world with black holes took roughly two generations to complete. Although nature might eventually decide the issues in the end, social prestige can still impose a wrong-headed conclusion for quite a while. Suppose the issues are extremely subtle and complex; the observations and measurements needed to resolve the issues are difficult to acquire; and there is no substantial theoretical context within which to work. If all of these statements are true, then a situation in which social factors prevail over nature's voice might last for some time. The postmodern critics can be right for a given historical period, even if nature does ultimately constrain our constructions in some profound manner.

§2. UNDERDETERMINATION AND IDEOLOGY

The Quine-Duhem Thesis Revisited

There is a variation of the constructivist argument, based on the Quine-Duhem thesis (see chapter 14), that would tend to blunt the point I just made. The basic idea of the Quine-Duhem thesis is that scientific theories are not uniquely determined by observations and experiments. A given set of results can be interpreted in a number of different ways. If the data aren't deciding the content of our theories, then we must (at least in part) be freely constructing them. Put differently: We don't discover scientific results, we invent those results. Since science is done by people working together in the scientific community, inventing results is a social process. While there is surely a certain amount of truth in this outlook, careless application of the idea leads to an extremely distorted vision of science. The stories of discovery related in Part I, taken collectively, reveal a much richer tapestry than the threadbare statement that science is a social construction. Science is a construction, but by no means is it an arbitrary construction. Our data may not uniquely determine our theories, but our data working together with our demand for logical coherence effectively whittles down the possibilities quite dramatically.

Two major problems plague the postmodernist use of the underdetermination argument. One problem is the blithe assumption that it's easy to come up with any number of alternative theories that explain a myriad of empirical observations. In fact, as anyone who has made the attempt will tell you, it's usually difficult to come up with even *one* satisfactory explanation. Since all of our explanations must interlock coherently, the Quine-Duhem thesis fades in relevance. The second problem goes deeper in its implications. A really good scientific theory gives us back much more than we put into it. A theory devised to solve one problem turns out to have the power to solve hundreds of unanticipated problems. An explanation for what is now known might also explain new discoveries made decades from now, discoveries that no one could ever have foreseen. This unexpected power of theories to advance far beyond the boundaries within which they were invented is difficult to understand if such theories are merely arbitrary constructions. To me, this suggests that these theories are in some sense a mirror of reality. The claim here isn't that science gives us a complete and undistorted grasp of reality; or that science allows us to apprehend all possible levels of reality; or that science is the only valid approach to reality. However, scientific results are not arbitrary; we achieve an understanding of nature that is meaningful and genuine in the course of scientific investigation.

Theories, Metaphors, and Cultural Myths

We need to be very careful about the interpretations we draw from scientific results. The postmodern commentators on science make some valid criticisms, but I think these criticisms are often misdirected toward science itself rather than questionable interpretations that pretend to be science. For example, a scientist might proclaim that humanity has been dethroned from a central place in the cosmos. A scientist might make this statement, but it's not science. Such a statement is a metaphorical extension of the actual scientific facts and is highly charged with emotional overtones and value implications. Many cultural critics have castigated science for being soulless, mechanistic, atheist, capitalist, or what have you. Both scientists and critics of science have unfortunately sometimes confused scientific results with the value-laden interpretations that they overlay on these results. (A more extended discussion of this issue is found in chapter 11.) The problems involved in disentangling metaphors from theories can be subtle because the metaphors are often crucial in the process of constructing the theories. In addition, these metaphors become entrenched in the interpretations of what a theory means, especially in the larger culture. A clear example is evolutionary theory, which has a well-defined role in science (providing a coherent explanation of a set of observations), a broader role in both science and culture (providing, in Darwin's words, a "view of life"), and finally an extremely broad role in culture as a contemporary "creation myth." All of these roles might be legitimate on some grounds, but only the first is legitimate as science. Making these distinctions isn't easy in practice, however, because the different roles all operate simultaneously under the same name (evolution by natural selection).

Back to Politics

A major claim of the postmodern program is that the results of science merely reflect the political ideologies, economic interests, and cultural prejudices of scientists (and/or the elites who employ the scientists). I have argued that this statement, taken literally as a kind of universal assertion, is wrong. But if we back away from this extreme position and ask ourselves where a less rigid version might apply, we find a more fertile area for discourse. A number of postmodern scholars have done this sort of worthwhile analysis, and it's unfortunate that more extremist claims have dominated the discussion. For example, political and economic considerations have clearly played a large role in shaping the scientific research agenda (military, agricultural, and medical research are all examples). These extrascientific influences don't completely shape the research agenda, since the science itself drives research into unexpected directions, but their influence is undeniable. Similarly, there are real questions concerning gender discrimination in the sciences (even if the claim that science is a manifestation of "male thinking" is nonsense).

Finally, let's reconsider the situation where underdetermination is most important (complex phenomena, sparse data, and no fundamental theory). If deeply held values are also involved, then hidden ideological presuppositions might well dominate the discussion. As an example, consider the long-running debate over the question of environmental influences versus inherited traits in "determining intelligence." The quotation marks are there because it's not clear that causal determination is even the proper way to frame the issue and ask questions. Nor is it clear that intelligence is a single well-defined concept, much less an independent object of study. Vast social and economic resources might be redirected based on the outcome of this debate. The questions involved touch on deep underlying visions of what it means to be human. Given all of these circumstances, proponents of a particular (and definitive) position on this issue (the nature/nurture debate over intelligence) aren't basing their position on purely scientific grounds. If the postmodern critique draws attention to the murky foundation of issues like this, then I think the postmodernists have performed a valuable service. Perhaps the outcome of such a critique can isolate the genuinely scientific portions of the debate and the valid conclusions that can be drawn.

FOR FURTHER READING

Revolutions and Reconstructions in the Philosophy of Science, by Mary Hesse, Indiana University Press, 1980.

Construction and Constraint, edited by Ernan McMullin, University of Notre Dame Press, 1988.

Reconstructing Scientific Revolutions, by Paul Hoyningen-Huene, University of Chicago Press, 1993.

The Flight from Science and Reason, edited by P. R. Gross, N. Levitt, and M. W. Lewis, New York Academy of Sciences, 1996.

Introduction to the Philosophy of Science, by Robert Klee, Oxford University Press, 1997.

EPIGRAPH REFERENCE: William James, *Pragmatism*, The New American Library, 1974 (original edition 1907), p. 142.

PART IV

COMMON GROUND: SOME UNIFYING CONCEPTS

IN THE SCIENCES

Chapter 16

FLEAS AND GIANTS: SOME FASCINATING INSIGHTS
ABOUT AREA, VOLUME, AND SIZE

> The simplest plants, such as the green algae growing in
> stagnant water or on the bark of trees, are mere round cells.
> The higher plants increase their surface by putting out leaves
> and roots. Comparative anatomy is largely the story of the
> struggle to increase surface in proportion to volume.
> *(J. B. S. Haldane)*

WHAT DO THE FOLLOWING all have in common: mittens;
the chemical industry; your lungs; kindling wood for camp-
fires; and the low heating bills of rowhouses? Answer: all these
things, and many more, depend on the mathematical relationships between the surface area of an object, its volume, and the characteristic size
of that object. These relationships are relatively simple, but their implications aren't always obvious at first glance. We'll explore these implications in some detail, and our reward will be an idea that is both simple and
powerful, an idea that explains many seemingly unrelated phenomena
in many different sciences both effortlessly and elegantly. A little effort
examining the purely geometric concepts in the beginning is more than
repaid later with a host of fascinating applications.

§1. Basic Ideas

Squares and Cubes

Look at the cube drawn in Figure 8. As you can see, the cube has six (6)
square faces and twelve (12) straight line edges. Every one of these 12
edges has the same length. Let's use the symbol L to designate this length.
Every face of the cube is then a square, made up of four equal sides of
length L. If you asked me "How big is the cube?" or "What is the size of
the cube?", I would answer that the size of the cube is L; in other words,
I would use the cube's edge length, or linear dimension, to characterize
the size of the cube.

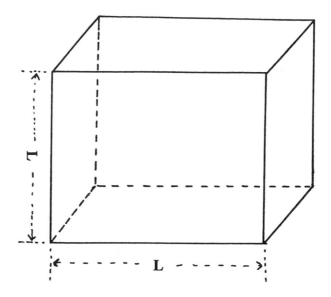

Figure 8. A cube of edge length L.

What is the total surface area of the cube? Each cube face is a square, and the area of a square is the length of the side multiplied by itself, that is, the length squared, which we write as L^2. The origin of the name L squared for $L \times L = L^2$ lies in this geometric relationship. Since there are 6 faces, the surface area of the cube is $6L^2$.

What is the volume of the cube? The volume of a cube is the length of a side multiplied by itself and then multiplied by itself again, that is, $L \times L \times L = L^3$. This is called L cubed, and once again is named for the geometric relationship. An illustration of these principles is shown in Figure 9 for the specific case of $L=2$. As you see, the area of a face is $L^2 = 4$, and the volume of the cube is $L^3 = 8$. Looking carefully at Figure 9 reveals exactly why the relationships between length, area, and volume are those we have specified. A key point here is that the surface area grows much more rapidly than the side length, and that the volume grows much more rapidly than the surface area. If we let A stand for the surface area of a face (recall $A_{cube} = 6A_{face}$), and V stand for the volume of the cube, we can summarize our relations so far as

$$A = L^2$$

and

$$V = L^3.$$

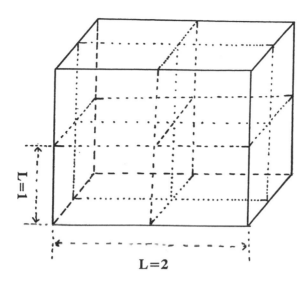

Figure 9. A cube of edge length $L = 2$, illustrating that the surface area of a cube face is L^2 (which is 4, in this case) and that the volume of a cube is L^3 (which is 8, in this case).

We see that for $L=1$, we have $A=1$ and $V=1$; for $L=2$, we have $A=4$ and $V=8$; for $L=3$, we have $A=9$ and $V=27$; and so on. The rapid growth of volume compared to area, and of area compared to length, is illustrated in Figure 10, which is a graph of V and A versus L.

Although I've suppressed the units so far to keep the discussion simpler and more focused, units are important. The unit of area is the square of the unit chosen for the length. If we measure length in feet (ft), then area is measured in square feet (ft²). If you purchase cloth for sewing, you might buy an amount of cloth measured in square yards. In the SI system of units (this is just the metric system; SI stands for the French words Système International), area might be in units of cm² or m². For volume, the unit is the cube of the length unit. For example, you might want to know how many cubic feet of water are in a gallon. Again, the SI units of volume might be cm³ or m³. In scientific work, the SI units are usually preferred, and I'll generally either use SI units or else suppress the units as I did before. As long as a consistent set of units (e.g., cm, cm², cm³) is used, the situation is fairly simple and we can suppress the units without affecting any of the relationships we've studied.

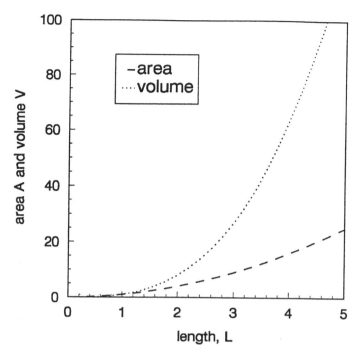

Figure 10. The surface area of a cube face and the volume of the cube plotted against the edge length of the cube, illustrating the rapid growth of the volume compared to the surface area, and the rapid growth of the surface area compared to the size of the cube.

The Crucial Point

We now come to the fascinating and surprising result that is the main point of this chapter. Because the volume is growing faster than the surface area as the length increases, *the ratio of surface area to volume decreases with increasing size.* Let's demonstrate this point explicitly for cubes of side length $L = 1$, 2, and 3 (again suppressing the units). The areas of the square faces are $A = 1$, 4, and 9, so the surface areas of the cubes are $A_{cube} = 6$, 24, and 54. The volumes of the cubes are $V = 1$, 8, and 27. The ratios of surface area to volume for these three cubes are then equal to 6/1, 24/8, and 54/27 (which in turn are equal to 6, 3, and 2). So, we see that as L increases, the amount of surface area per volume decreases. But we can turn this around to get a very important result: *the ratio of surface area to volume increases with decreasing size.* This key

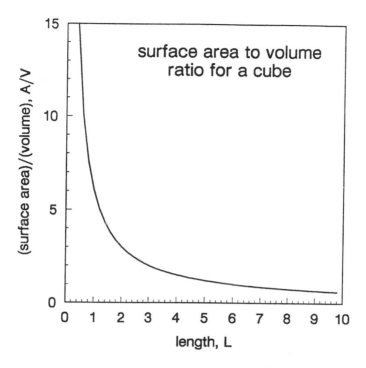

Figure 11. The ratio of the surface area of a cube to the volume of the cube, plotted against the edge length L of the cube. This graph illustrates the dramatic increase in this ratio as the size of the cube decreases.

relationship is shown graphically in Figure 11, where the (surface area)/(volume) for a cube is graphed against side length L.

Another way to see why these relationships are true is illustrated by looking back at Figure 9. Imagine that this cube with $L=2$ (and thus $V=8$) is sliced up into eight (8) smaller cubes of $L=1$ ($V=1$) by cutting through the dotted lines. Each of the new smaller cubes has three faces that were part of the surface area of the original large cube, but each also has three new faces that were part of the interior of the large cube, not part of its surface area. The same total volume now has twice as much surface area in the form of eight small cubes instead of one large cube. This same volume, as you can see, would have even more surface area if each of the small cubes was again sliced up into even smaller cubes of $L=1/2$ (we'd have 64 cubes and quadruple the original surface area). We can draw the following conclusions: A smaller cube has more surface area per unit volume than a larger cube. A given volume exposes more and more surface area as it is divided into smaller and smaller cubes. Why are these conclusions of any interest to science? Before we tackle this question, let's show that our conclusions are not restricted to cubes. Meanwhile, you can get a sense of the answer by thinking about a more specific question, namely: What would you rather use to cool down a drink quickly, one large ice cube or eight smaller ice cubes having the same total ice volume?

Circles and Spheres

We now turn our attention from the square to the circle (in two dimensions) and from the cube to the sphere (in three dimensions). What length can we use to characterize the size of a circle? The radius of the circle is an obvious choice (or the diameter, which is just double the radius). The area of a circle (as you may already know) is proportional to the square of the radius. If we let R stand for the radius of a circle, the area is given by the formula

$$A_{circle} = \pi R^2,$$

where π (the Greek letter pi) is a constant equal to the ratio of the circle's circumference to its diameter. The important point here is that the area is proportional to the square of the characteristic size (R^2), just as was true of the square itself (L^2). The surface area of a sphere is a little more complicated than that of a cube. We can't just add up some circle areas to get the area of a sphere. In the end, though, the formula for the surface area of a sphere turns out to be simple:

$$A = 4\pi R^2,$$

where R in this case is the radius of the sphere. Once again, the crucial point of interest is that the surface area of the sphere increases as the square of the sphere's linear dimension (radius).

The volume of a sphere is also difficult to derive but easy enough to write down. Again let R stand for the radius of the sphere. The volume is given by the formula

$$V = (4/3)\pi R^3.$$

Yet again, the interesting thing from our point of view is that, for a sphere, the volume is proportional to the cube of the radius. As the radius increases, the volume of the sphere will grow faster than the surface area of the sphere. Why? Because the volume and surface area of a sphere depend on the radius in the same way that the volume and surface area of a cube depend on the edge length. (Of course the proportionality constants are different.) The reasoning we applied to the case of the cube, and the conclusions we came to, are every bit as valid for the sphere. In particular, a large sphere that is subdivided into many small spheres will then have much more surface area.

We could demonstrate these facts with numbers and with graphs, as we did for the cube, but let's instead try a different technique. We're interested in the ratio of the surface area to the volume, A/V. If you replace A and V in this ratio with our formulas for a sphere, you'll find that $A/V = 3/R$. In other words, the surface area per unit volume is proportional to $1/R$. As R increases (i.e., the sphere gets bigger), the amount of surface area per volume decreases; as R decreases (the sphere gets smaller), the amount of surface area per volume increases. As we said, the sphere has the same properties as the cube in this respect.

Other Shapes

A variety of well-defined geometric shapes (like cones and pyramids) have known formulas for their surface areas and volumes. In all cases, the area is proportional to the square of some characteristic length, and the volume is proportional to the cube of this length. Strictly speaking, this is only true if all the relevant lengths (such as the height of a cone and the radius of its base) are equal; only the cube and sphere have a single length that completely characterizes the size. None of our conclusions are affected by this point, however, as you'll see next.

Suppose the shape is irregular, like a piece of fruit, a rock, or an animal. In these cases, there are no simple formulas for the volume and surface area. But even for such irregular shapes, the surface area increases as the square of some characteristic length, while the volume increases as the cube of this length. (The preceding examples have hopefully convinced you that this must be so. Thinking about the units of volume and of surface area will give you some insight into why.) The characteristic length isn't precisely defined for irregular shapes, but it doesn't need to be precisely defined either. We aren't interested in precise values for the surface area or volume; we are only interested in the consequences of how A and V depend on the size of the object. We'll use the symbol ℓ to stand for this ill-defined length, which characterizes the size of the object. The volume is then proportional to ℓ^3 and the surface area is proportional to ℓ^2. But the proportionality constants are not well defined and will vary depending on the exact shape of the object. (A useful way to express these relationships is $A \sim \ell^2$ and $V \sim \ell^3$ where the symbol $\sim$ is read "goes as" or "scales with" and has the meaning we have just given.) The important point is that these irregular shapes have roughly the same properties as the cube and the sphere, properties with which we are by now quite familiar.

The volume always grows more rapidly than the surface area as the size of an object increases, regardless of its shape. Smaller things have more surface area per unit volume than bigger things do. Suppose a large irregularly shaped object, like a rock, is ground up into many small irregularly shaped objects, like sand grains. The total volume of the sand is the same as the volume of the rock. But the sand has a vastly greater amount of surface area than the rock had. An elephant has a greater surface area than a mouse has, but the mouse has a much greater surface area *for its volume* than the elephant has. In other words, if we collect enough mice together so that their total volume is the same as the elephant's, the total surface area of the collected mice will be much greater than that of the elephant.

§2. Applications in Science and in Life

Of Mice and Mittens

So what difference does it make to the mouse whether it has more surface area per unit volume than an elephant? Actually, it makes a lot of difference. Animals lose body heat through their skin, that is, their surface area. Because heat is lost through the surface area of the skin, the mouse loses heat much faster than the elephant, relative to body weight. Since the heat comes from the food eaten by the animal, mice need to eat much more (again relative to body weight) than elephants. Another example involving body heat is one that I mentioned in the first sentence of the chapter, namely mittens. Why are mittens more effective than gloves to keep your hands warm? You lose your body heat through your exposed surface area, just like the mouse. Your fingers are relatively small parts of your body, so they expose relatively large amounts of surface area for their volume. A glove that wraps around each individual finger also exposes a lot of surface area for the volume of the hand, whereas a mitten is one large wrapping, which thereby has a smaller surface area exposed. Less heat is lost because the mitten is larger than the finger of a glove, and so it has (as we've learned) less surface area per volume.

Strength and Weight

A flea can jump about two feet into the air. You may have read or heard that if a flea were as big as a human, the flea could jump thousands of feet high. This statement is based on a simple proportion between the size of the flea (compared to a human) and its jumping ability; the size and the jump are both scaled up equally to reach this conclusion. But this conclusion isn't correct, because a flea's strength, which determines its jumping ability, increases in proportion to the cross-sectional area of its legs. The flea's weight, on the other hand, which it must propel upward, increases in proportion to the volume of the flea. Based on the relationships we've learned, the flea's weight will increase by hundreds of times more than its strength if the flea becomes hundreds of times bigger. If a flea were as big as a human, it could *not* jump thousands of feet into the air. In fact, if a flea were that big, it might not even be able to stand up. The same fate would befall the legendary giants found in folktales and myths. The ability of leg bones to carry weight will scale upward with the cross-sectional area of the bones, whereas the weight the bones must carry will scale upward with the volume of the giant. If the giant is 5 times as tall as a human, the bones of the giant are 25 times as strong; but these bones must carry 125 times as much weight. A large enough giant would not be able to walk, his legs having broken under his own weight. Despite my somewhat whimsical examples, these principles actually operate in real life in ways that you can see. Look at the body and legs of a deer, and then look at the body and legs of an elephant. The legs of the elephant are much thicker and sturdier, compared to the size of its body, than the legs of the deer. The very largest creatures on earth, the whales, are sea creatures that don't need to hold themselves up with legs. Small creatures, like insects and spiders, often have very thin legs.

Making Surface

A given volume of material has the least surface area when it is one object. Turning this volume into many smaller objects, by dividing it up, will increase the surface area. The smaller these objects are (which means the greater their number), the more surface area we create. This principle is exploited in many different ways, both by nature and by humans. A wonderful example is the biological cell. A cell gets its nutrition through the cell membrane. So, the ability of a cell to supply its needs is proportional to the cell's surface area. But the nutritional needs of the cell are proportional to the *volume* of the cell. As a cell gets bigger, its need for nutrition grows faster than its ability to supply the need. This fact (resulting from the mathematical relationships of volume and area to size) imposes limits on the maximum size that a cell can have. Single-celled creatures are therefore always small. But of course there are large creatures in nature. Nature has been able to sidestep this limit on size by means of multicellular organisms. The size of the cells remains small, but there are many of them.

A technological example of these principles is found in the operation of chemical catalysts. A catalyst is a substance that encourages (promotes) a chemical reaction, but is not itself changed by the reaction. A catalyst works by bringing the reacting molecules onto its surface, where the molecules find each other and undergo their reaction. The new molecule that results from the reaction then leaves the surface, making room for more reactants to start the process again. Clearly, based on this description, the effectiveness of a catalyst increases if it has more surface area. The surface area of a catalyst is increased by making it in the form of highly divided particles, like a powder. Such catalysts are essential to the activities of the chemical industry. We find a similar application in the operation of filters, like the kind used to purify water. In this case, however, the impurities sticking to the surface of the filter material stay there. Once again, more surface area is desirable for improving the effectiveness of the filter, and more surface area is created by making the filter material in the form of small particles. The surface area of a material can also be increased by making the material highly porous. This is actually similar to having the material highly dispersed into a powder. (Think of the powder as being compressed with the small particles glued together, and you have a porous material.) Groundwater is purified by seeping through porous rocks on its way to the underground water table.

Turning again to an example from biology, consider the operation of your lungs. The job of the lungs is to deliver oxygen from the air to your bloodstream, where it is carried by the red blood cells. To get from the air to your blood, the oxygen must move across a surface. The rate at which the blood can be oxygenated, then, is limited by the amount of surface area available for this transfer. If your lungs were just hollow like balloons, the amount of surface area they would have (given their volume, which is your chest cavity) would be many hundreds of times too small to supply the amount of oxygen you need. You would soon be dead. So how can the lungs do their job? The total volume of each lung is subdivided into many tiny air sacs, called alveoli, each of which has a much greater surface area per unit volume than one large air sac the size of the lung would have. The tubes bringing air to the lungs keep branching into smaller and smaller tubes until the microscopic alveoli are reached. The walls of the alveoli have blood capillaries in them, and this is where the exchange of oxygen for carbon dioxide takes place. The vast number of alveoli into which the volume of the lungs is divided contain a vast amount of surface area in which this gas exchange takes place. In a sense, the mathematical principles we studied in §1 are responsible for our being able to breath.

These principles also operate in geology. For example, erosion is a process that basically occurs at a surface, since this is where the wind or water actually attacks the eroding substance. As the erosion advances and the substance begins to break down, more surface area is exposed and the process will accelerate. Corrosion processes attack metals in a similar way for similar reasons. Corrosion often starts at a small crack or fissure, where the surface area is greater. As the metal corrodes, more surface area is exposed. You may have noticed a small pointlike area of rust on your car, which stayed about the same size for a while, but became bigger faster once it started to grow. Another example from geology is the movement and retention of water in porous rock and soil systems. Finer soil particles present more surface area and thus hold more water (the water sticks to the surfaces) than coarse sandy soils.

Examples from Everyday Life

The same processes that occur in erosion and corrosion also determine how quickly salt and sugar dissolve in water. Large pieces take a longer time to dissolve because they have less surface area per volume than finely ground powders and the dissolving process occurs at the surface. Another example from the kitchen is the technique of grinding spices to release their flavors more effectively (a whole peppercorn, because it has less surface area, has much less effect than a ground peppercorn). The same is true of ground coffee. An ounce of chocolate chips will melt much faster than a one-ounce brick of chocolate. If you have ever built a fire, you know that you can't just light a match under a log. You start with paper, use the paper to get kindling wood (small pieces) burning, then add moderate-sized pieces of wood. When the fairly large pieces of wood are burning well, then you can put big logs on the fire. Fire is a chemical reaction between the wood and oxygen, and this reaction takes place at the surface. Exposing more surface area by having smaller pieces of wood makes it easier to get the fire started. The fire won't last long, though, because the volume of these small pieces won't provide fuel for long. After the fire is well started, then you add the big pieces of wood, which have a lot of volume (but not much surface) in order to keep the fire going.

Household examples of these principles are easy to find. A basket of wet laundry will mildew before it dries because drying occurs at the surface and the bundled cloth exposes little surface area for its volume. The laundry needs to be hung up (which exposes a much greater surface area) in order for it to dry. Snow that has been piled up by drifting or plowing doesn't melt until long after the rest of the snow has melted. The large piles have less surface area for their volume, and the snow absorbs heat through its surface. Though you can't see it directly, another example is the carburetor of an automobile, which turns the gasoline into a fine mist to expose more surface area for ignition. A building loses heat through its outside walls, so the rate of heat loss in the winter is proportional to the surface area of the building. The amount of heat needed to warm the inside of the building is proportional to the volume. A large apartment building, or a set of rowhouses, has less surface area per volume than a number of detached houses having the same total volume. While the heating bill of the apartment building is probably larger than that of any one house, the heating bill per dwelling unit will certainly be smaller for the apartment building than the heating bill for the average house, all other things being equal. In an apartment, most of your walls are inside walls through which you are not losing any heat. For our last example, consider the prices of different pizza sizes (small, medium, and large). The cost of a pizza is often roughly proportional to its diameter (i.e., a linear dimension), but the amount of pizza you eat is better measured by its area (which is proportional to the square of this linear size). Based on our familiar mathematical reasoning, you will get more for your money by buying a large instead of a small.

Recap

Scientists regard an idea as powerful if the idea ties together a large number of seemingly disparate phenomena in a coherent way, if many separate observations can be explained by a single underlying cause. If the idea is easy to understand, if it is simple as well as powerful, then the idea is especially worthwhile. The relationships we've explored in this chapter, namely,

$$A \sim \ell^2$$

and

$$V \sim \ell^3,$$

qualify as both simple and powerful. Their mathematical simplicity translates into an intuitive sense that surface area per unit volume shrinks rapidly with increasing size. Their implications for the behavior of things in the world are highly important, not only in all of the natural sciences but in our everyday lives as well.

For Further Reading

"On Being the Right Size," by J. B. S. Haldane, reprinted in *The World of Mathematics*, edited by J. R. Newman, Simon and Schuster, 1956, p. 952.

"On Magnitude," by D'Arcy Wentworth Thompson, reprinted in *The World of Mathematics*, edited by J. R. Newman, Simon and Schuster, 1956, p. 1001.

"Scaling," in *Conceptual Physics*, by P. G. Hewitt, HarperCollins, 1993, p. 201.

Epigraph reference: J. B. S. Haldane, in *The World of Mathematics*, edited by J. R. Newman, Simon & Schuster, 1956, p. 954.

Chapter 17

THE EDGE OF THE ABYSS: ORDER AND DISORDER IN THE UNIVERSE

> What was visible was chaos, irregularly streaming
> bits of liquid, random motions. . . . Then suddenly . . .
> crystallization began . . . resplendent with colorful
> order and geometric beauty.
> *(Henry Margenau)*

> Movement overcomes cold.
> Stillness overcomes heat.
> Stillness and tranquillity set things in order in the universe.
> *(Lao Tsu)*

> The gap between "simple" and "complex," between
> "disorder" and "order," is much narrower than
> previously thought.
> *(G. Nicolis & I. Prigogine)*

> This order, the same for all things, no god or man has made.
> *(Heraclitus)*

IN THE CREATION MYTHS of many cultures, the divine powers engage in a mighty struggle to impose form on the primordial forces of chaos. The world is a continual struggle to maintain form and order, which forever hovers close to the edge of the abyss. With this mythic and poetic backdrop to remind us of the broad issues underlying our investigation, let's take a look at how modern science approaches the age-old question of order and disorder in the universe. While the scientific worldview is a bit less poetic, some remarkable insights have emerged from the study of these questions. The last several decades in particular have seen a resurgence of interest, accompanied by a new understanding, which many people consider a scientific revolution. Before looking at these revolutionary new results, however, let's start by considering the older (and no less interesting) insights about order and disorder that are rooted in the nineteenth century studies of heat.

§1. Disorder from Order—The Second Law of Thermodynamics

Heat and Motion

Heat had once been pictured as some sort of subtle fluid, but by the middle of the nineteenth century we understood that heat is a form of motion: the microscopic motion of the atoms making up an object. These moving atoms have energy (called kinetic energy, meaning energy of motion), just as a moving baseball or a moving planet (or anything else that moves) has energy. But you can see the motion of a baseball, whereas the motions of the atoms are invisible, and in many different random directions. Although we can't see this random microscopic motion, however, we can feel its effects as heat. To convince yourself dramatically of the equivalence of motion and heat, feel a piece of metal that you have just been hacksawing or drilling for a while. It's pretty darned hot! This observation is exactly what led Count Rumford, based on his classic cannonboring experiments, to conclude that heat is a form of motion. What does all this discussion about heat have to do with order and disorder? Well, the thermal motion of the atoms is certainly more random and disordered than the back-and-forth motion of a hacksaw or the spinning motion of a drill. This simple observation marks the beginning of an elaborate theoretical development, which we now call thermodynamics and statistical mechanics, at the hands of Carnot, Joule, Clausius, Kelvin, Helmholtz, Maxwell, Gibbs, and Boltzmann. Let's trace this development in a little more detail, and then we'll return for a deeper look at our major theme of order and disorder.

The First Law

Much of the interest in these problems during the nineteenth century was due to the practical problems of the industrial revolution. The main source of power at that time was the steam engine. So far, we've talked about motion (like drilling) turning into heat; a steam engine does the reverse of this process, turning heat into motion (burning coal to run a train locomotive, for example). Since the heat and the motion are both forms of energy, these examples are specific cases of a more general process: the transformation of energy from one form into another. Energy transformation processes occur all the time in nature. Energy in sunlight, for example, gets turned into energy stored in chemical bonds during photosynthesis in plants. When we eat the plants, this stored chemical energy is turned into energy of motion by our muscles, into heat to keep our warm-blooded bodies warm, and into the energy needed by our cells to stay alive. Millions of years ago, some of the plants were buried and slowly became coal or oil. The energy stored in these fossil fuels is turned into heat when they're burned; this heat energy is turned into motion by steam turbines, automobile engines, and so on. The steam turbines often turn electrical generators, which transform the energy of motion into electrical energy. When you switch on an electric lamp, you may be using some of this primordial energy, which came to earth in ancient sunlight.

These examples hint at the vast range and importance of energy transformations in nature. The scientists of the nineteenth century discovered a remarkable fact about these transformations; although the energy might change its form in myriad ways, the total *amount* of the energy is always the same before and after the change. No energy is ever lost or gained in any natural process. This statement is the famous law of conservation of energy. Careful experiments conducted by James Joule were the first convincing demonstration of this law, and no exceptions have ever been discovered since then. Although the applications of this principle (conservation of energy) extend far beyond its origins in the study of heat, the original name is also still used: the first law of thermodynamics. This law tells us that energy can be neither created nor destroyed, but only changed from one form into another.

Conservation of energy is one of the most general and important principles in the sciences. Chemical reactions, biological metabolism, and engineering design practice are all governed by this principle. It forbids, once and for all, the possibility of a perpetual motion machine that does work for us without requiring fuel (see chapter 12). In physics, energy conservation survived the revolutionary changes in worldview wrought by relativity and quantum mechanics; space, time, and determinism all changed irrevocably, but energy conservation still holds true. In fact, when physicists were faced with an apparent breakdown of energy conservation in a subatomic reaction, they postulated the existence of a new particle with properties that made it difficult to observe yet allowed it to conserve energy in the reaction. Sure enough, several decades later this particle (the neutrino) was experimentally verified to really exist.

The Second Law

But the first law doesn't tell us everything we need to know. A process that violates conservation of energy is impossible, but many processes that conserve energy perfectly well are *still* impossible. For example, a golf ball could turn much of its thermal energy into motion, lowering its temperature and jumping spontaneously into the air, without violating the first law. But no golf ball has ever done this. Why not? To answer this question, we must distinguish between energy that we can use to do useful work (like move an object), and energy that is not usable in this way. For example, if I expend some of my energy lifting a box, then I can get this energy back by dropping the box (which acquires energy of motion as it falls). But if I expend my energy pushing a box across the floor, I've lost this energy in the sense that the box won't move again on its own. My energy has been dissipated by friction into an unusable form. The energy is conserved (has not been lost to the universe), but cannot be recovered. Such dissipated energy often ends up as heat (friction between the drill and metal causes the heat I mentioned previously). We can extract useful work from heat; that's what a steam engine does. But we can't turn *all* of the heat into work. Motivated by practical interest in steam engines, Sadi Carnot investigated carefully the question of just how much usable energy can be extracted from heat. We now call Carnot's answer to this question the second law of thermodynamics. The second law is what prevents our golf ball from spontaneously jumping into the air.

To understand the second law, we must first introduce the concept of entropy. Entropy is an abstract quantity. You can't see it or feel it. The change in the entropy of an object is defined as the ratio of heat change (gained or lost) to temperature. Admittedly, the interpretation of entropy, defined in this way, is a little murky (we'll clear it up later). But Rudolf Clausius realized that the quantity that he termed entropy had a remarkable property: it never spontaneously decreased in any physical process. This property of entropy is the essence of the second law. As Clausius phrased it, "The entropy of the world tends toward a maximum." A simple example will help illustrate the meaning of the second law. What happens when a hot (higher temperature) block of metal touches a cool (lower temperature) block of metal? The hotter metal cools, of course, and the cooler metal warms, until both are at the same temperature. Heat energy flows from the warmer to the cooler. While the hot object loses entropy and the cool object gains entropy, the cool object gains more entropy than the hot object loses based on our definition of entropy. The total entropy of the system (i.e., both blocks together) increases, in accordance with the second law. A cake taken out of the oven cools in order to increase the total entropy of the world. The trivial-sounding observation that "heat never spontaneously flows from a cooler object to a warmer object" can actually be taken as an alternative statement of the second law.

Heat does flow from higher temperatures to lower temperatures, and when it does, we can extract some useful work from this heat energy. The maximum amount of work we can obtain corresponds to a total entropy change of zero (an ideal heat engine); to get more work would decrease the entropy and violate the second law. Any real-life engine actually increases entropy as it operates, and produces even less usable work. These considerations are obviously relevant to engineering designs. In addition, the second law governs whether chemical reactions occur, determines the distribution of plants and animals in a food chain, is centrally important to virtually every science, and even prohibits the existence of perpetual motion machines that don't do work and so violate the first law. But what does any of this have to do with order and disorder?

Entropy and Information

The answer to this last question was discovered by Ludwig Boltzmann, working in Vienna near the end of the nineteenth century. To understand Boltzmann's conception of entropy, we must first look in general at the project that Boltzmann (and also Willard Gibbs at Yale) undertook to accomplish. Thermodynamics does not try to tell us anything about the motions of the atoms in a substance. Heat is simply taken to be another form of energy, and the second law is a compact distillation of a vast array of observations for which there is no further explanation. Boltzmann and Gibbs created a microscopic theory (i.e., a theory dealing with the motions of atoms) and explained the thermodynamic laws in terms of these microscopic motions. This theory is called statistical mechanics. The word "statistical" appears in this name for an interesting reason. We are now trying to understand the behavior of a substance by looking at the behavior of its atoms. But there are an unimaginably large number of these atoms. We can't possibly know how each of the atoms behaves. How then can we understand the properties of a substance based on our understanding of its atoms? Boltzmann and Gibbs accomplished this feat by using probability and statistics. We can't know the behavior of each individual atom, but we *can* predict the net behavior of a large number of atoms statistically. (An analogy might be flipping coins; we can't predict whether any single coin will be heads or tails, but the fraction of a large number of flips coming up heads is predictable, namely, one-half.) The extremely large number of atoms works to our advantage now, since statistical results become more precise for larger samples. The results of thermodynamics, based on observation, can all be explained by this application of probability and statistics to motions at the atomic level. We have, in effect, reduced thermodynamics to statistical mechanics (see chapter 14).

Boltzmann discovered that entropy, in this theory, is a measure of the number of states available to the system. What does this mean? Imagine a deck of cards with all four suits separated and arranged from ace to king and the suits stacked in order (e.g., hearts, clubs, spades, diamonds). There is only one way to do this. This system has only one available state. The entropy of this system is at a minimum. Suppose we keep the suits separated and arranged, but now stack the suits in any random order. There are 24 ways to do this (count them), so the system now has 24 available states. If we shuffle the hearts so that they are arranged in random order, they can wind up in any of over 6 billion possibilities; the number of available states for this deck of cards is getting very high. If we simply shuffle the entire deck, then the number of possible arrangements (a *very* large number) is at a maximum. Since this is also the number of available states, the entropy now is at a maximum. As this example indicates, entropy has a very intuitive meaning. *Entropy is a measure of the disorder in a system.* The entropy is least for the completely well-ordered deck, the entropy increases with increasing randomness, and the entropy is at a maximum for a completely shuffled deck having no order. The second law now takes on a new significance. The second law tells us that *the disorder of the universe is increasing.* The natural tendency for a system is to become more disordered. Imposing order requires some effort (or, to be more precise, some energy). We can now understand why the second law is true. A highly ordered state is less probable than a disordered state because the number of choices for the system decreases as it becomes ordered. Heat flowing spontaneously from a cooler place to a hotter place is improbable for the same reason that shuffling a deck of cards into four arranged suits is improbable. For the disordered heat of a golf ball to become ordered energy of motion and make the ball jump into the air is so improbable that we'll never see it happen during the age of the universe. It's effectively impossible. All violations of the second law are like this: merely improbable, but so improbable that we declare them impossible. (A monkey pecking at a typewriter is far more likely to write every book in the Library of Congress than our golf ball is likely to jump in the air.)

This broader interpretation of the second law has wide application in nature. We can use it to explain why a drop of ink spreads throughout a glass of water, why dead plants and animals decay, and why machines are constantly in need of maintenance. A major new set of insights were worked out more recently, linking the concept of entropy with that of information. Entropy is a measure of disorder in a system (what we don't know), while information (what we do know) is in some sense a measure of the order that exists. Information, in this picture, then becomes a kind of negative entropy. These ideas (worked out in what we now call information theory) have proven to be exceedingly fertile in applications to modern telecommunications systems, including the now-famous internet. Another fascinating application of the second law is to an old problem, the arrow of time. We clearly perceive a directionality in time; the past is what has already happened, and the future has yet to happen. This trivial-sounding concept has a problem, from a scientific point of view: nothing in the equations governing the behavior of matter distinguishes between the past and the future. The world could just as well be running backward. This premise is surely absurd, and yet it demands an explanation. The second law provides such an explanation, because the disorder in the world must inexorably increase with time. In this view, the future is the time direction in which the disorder becomes greater. Increasing disorder itself imparts a direction to time's arrow. While this idea certainly isn't the last word on issues regarding time, it has profoundly influenced the discussion.

The ultimate significance of the second law is not entirely clear. A common misinterpretation is that the second law prohibits the formation of order, but this isn't true. Living things create and maintain highly ordered structures by consuming energy and creating greater entropy in the rest of the world. A more homely example is the refrigerator, which makes heat flow from a cooler place to a warmer place by consuming energy and discharging entropy. We can always create islands of local order at the expense of greater disorder elsewhere. Only a so-called isolated system can never decrease its entropy. Of course, the ultimate isolated system is the entire universe, and pessimistic writers have taken a nihilistic message from the second law: everything is headed toward total disorder. Extrapolating our limited knowledge and experience to the entire universe is always dangerous, however, and there are still questions about the cosmic significance of the second law. What we *do* know about the central role of the second law in chemistry, physics, biology, geology, engineering, and information theory is already important enough.

§2. Order from Disorder—Open Systems and Emergent Properties

An isolated system is also sometimes called a closed system, in contrast to an open system that can exchange energy, material, and information with the rest of the world. For many years, the study of thermodynamics was essentially restricted to the study of closed systems at equilibrium. (By equilibrium we mean the final steady-state condition of a system; for example, when our hot and cold metal blocks arrive at the same temperature, they have achieved equilibrium.) The entropy of a closed system is maximum at equilibrium. Closed systems at or near equilibrium are the easiest to understand, but we have discovered more recently that such systems are by no means always the most interesting. In the last few decades of the twentieth century, the study of open systems that are far from equilibrium has revealed a wealth of new and revolutionary insights. We have discovered that order can spontaneously arise in such systems, a phenomenon known as self-organization. Self-organizing systems occur in living things, chemical reactions, moving fluids, computer models, and social organizations. We'll look at several examples of such systems, and we'll see how new properties (called emergent properties) can arise.

Some Simple Examples: Convection Rolls and Chemical Clocks

Let's start with a simple (but historically important) example: convection cells in a heated liquid. Imagine a wide, shallow container filled with water. The container has flat metal plates at the bottom and top, with water filling all the space in between. If the bottom plate is warmer than the top plate, heat will flow in order to equalize the temperatures (trying to achieve equilibrium). But if we continuously supply heat to the bottom plate, we thwart the attempt to achieve equilibrium. A steady-state temperature difference (called a gradient) is established instead. As we make the bottom plate hotter and hotter, the gradient increases, and we drive the system farther away from equilibrium. At a certain critical temperature difference, a remarkable thing happens. The system undergoes an abrupt transition and a set of rolling convection cells form. In these convection cells, warmer water from the bottom rises up (where it cools), driving cooler water down to the bottom (where it warms). The water moves around continuously, much like a spinning cylinder (these movements are sometimes called convection rolls). Each cell can roll either clockwise or counterclockwise, with the direction alternating from one cell to the next. This seemingly simple motion is quite remarkable because the entire system organizes itself into this pattern all at once. Order has arisen spontaneously from the disordered motion of heat.

No violation of the second law has occurred here, because we are constantly supplying energy to the system (which is far from equilibrium). The simplicity of the experimental set-up (called a Bénard cell) allows us to understand in some (mathematical) detail how and why the order arises in this case. We can vary the geometry and carefully control the temperature in order to make predictions that test our understanding. The basic ideas inherent in this simple experiment are also found in the titanic forces of nature. The great forces that slowly move continents are due to convection cells in the earth's mantle (see chapter 2). The circulating air of a hurricane or tornado and the vast stable ocean currents (like the Gulf Stream) are similar examples of self-organized convection systems in fluids.

We now turn to a completely different kind of system, namely a collection of chemicals and their reactions. Even a simple system, with just a few chemicals, can exhibit interesting properties of self-organization. The key requirement needed for this to occur is that some chemicals must play two distinctly different roles in the reactions. These chemicals need to be both reactants (taking part as either input sources or output products) and also catalysts (agents that speed up a reaction without taking part). This property is sometimes called autocatalysis. In addition, we must have an open system in which source materials, products, and energy can flow in and out. A famous example of such self-organization is the Belousov-Zhabotinski reaction, the so-called chemical clock. The chemicals involved are not exotic (cerium sulfate, malonic acid, and potassium bromate), but one of the intermediate products (bromous acid) catalyzes its own formation. The inflowing reactants drive the system from equilibrium, and (under the proper flow conditions) the reaction starts to periodically oscillate. Because two of the products have different colors, you can actually see the reaction oscillate as the solution periodically changes back and forth from colorless to yellow. The changes occur at highly regular time intervals, hence the name "chemical clock." If the solution isn't stirred, then differences can develop in space as well as time. Waves of color move through the solution, sometimes forming beautiful spiral patterns. All of this order unfolds naturally from the process itself, emerging from the microscopic disorder of the inflowing reactants.

In the previous two examples, we've only looked at self-organized behavior in simple systems. Highly complex systems, having many components interacting in a variety of ways, are actually more prone to self-organization than simple systems. Complex systems, in fact, are where the ideas of self-organization and emergence really come into their own. We'll soon look at highly complex autocatalytic chemical systems and how these systems relate to life. But first, let's look at a complex system made up of very simple pieces: Boolean networks.

Complex Networks

Each component of a Boolean network operates according to the rules formulated by George Boole over a century ago. These rules, which are the foundation of mathematical logic, operate on binary variables (which is a fancy way to say that the variable has only two possible values). We might call these two values by several names: true and false (if we're thinking about logic); on and off (if we're thinking about switches); zero and one (if we're thinking about binary, or base two, numbers). Since electronic devices can be built that have on/off states, their outputs can be used to represent either binary numbers or logical decisions. Electronic devices of this type, operating according to the rules of Boolean logic, are called logic gates. Such devices are at the heart of every digital computer.

The basic rules of Boolean logic are not complicated. Each component has multiple inputs and a single output. These inputs and outputs specify binary states (for example, let's call these states 0 and 1). The values of the input states then determine the output state, according to the rule we choose. An important Boolean rule is the *AND* operator, which specifies that the output will only be 1 if all of the inputs are 1. Supposing we have two inputs, then inputs 1 *AND* 1 give an output of 1. Inputs of 1 *AND* 0, 0 *AND* 1, and 0 *AND* 0 all give an output of 0. A different Boolean rule is the *OR* operator, which gives a 1 if *any* of the inputs are 1. In this case, we have 1 *OR* 1 gives 1, 1 *OR* 0 gives 1, 0 *OR* 1 gives 1, and 0 *OR* 0 gives 0. There are also other Boolean rules, but the definitions of the *AND* and *OR* are enough to get across the essential idea. We can string such components together in a complicated fashion to build up complex systems. As long as the inputs are kept completely separate from the outputs, the results are quite predictable and well-defined. If we mix the inputs and outputs together, however, letting the outputs of some components be the inputs of the others in a set of feedback loops (see chapter 21), then we have a complex system whose total state will be unpredictable. This arrangement is a Boolean network. The total state of the system is simply the complete specification of all the individual components' states. For example, a five-component system with a total state of {10100} means that the first and third components are in state 1, while the rest are in state 0. A five-component system has 32 possible states that it might be in. For a system with thousands of components, the number of states it might be in becomes astronomically large.

A Boolean network with thousands of components may have a highly disordered state. If the output of each component serves as the input for a large number of other components, the total state of the system cycles aimlessly among virtually every possibility. Little or no order is formed in such a system. Using the language of the previous section, the entropy of the system is near its maximum value. In the extreme case where every output is connected to the input of every component, the state is totally random. Every possible state is equally probable, and the entropy is as high as possible. If, however, we connect each output to the input of only two other components, we see a very different behavior. In this case, the system eliminates possible states at an astonishing rate. Instead of the astronomically large number I just mentioned, the system narrows its choices down to fewer than a hundred states. This drastic winnowing of possibilities is a spontaneous action of the system itself, and it happens for virtually any random interconnections made in the wiring. In other words, the self-organization doesn't depend on any details of the system design, but only on the restriction to two (or fewer) inputs to each member. (Of course, these wiring details do determine, in some unpredictable way, which of the states the system ultimately chooses to keep.) So, we see here the *spontaneous emergence of order from randomness* in a case where the individual rules for each component are simple, but the entire system is highly complex. Such emergence is not inherent in the rules, resulting instead from the operation of the entire system as a whole.

Somewhat similar systems, studied by many people, go by the name "cellular automata." In this case, each component is connected only to its nearest neighbors in the network. The number of states for each component might be more than two, and the rules governing the state of each component are more flexible than the Boolean rules. A famous example is the so-called Game of Life (invented by John H. Conway), in which each component has two states: alive and dead. The state of any component is determined by the states of its neighbors (i.e., by how many of its neighbors are alive or dead). The rule is simple enough, but the complexity arises from two facts: all of the neighbors are influencing *each other* at the same time, and the whole system *evolves* from one step to the next. Astoundingly, these simple rules can generate patterns that grow and reproduce themselves.

Emergence and Life

We've looked at self-organization in autocatalytic chemical reactions and in complex networks of interacting components. Suppose we now combine these cases by making the interacting components of the complex network a large set of chemicals that can react and autocatalyze to create the feedback interconnections. Based on our experience with Boolean networks and autocatalytic chemistry, we might expect such a system to self-organize into an ordered state of some sort. If these chemicals are amino acids, carbohydrates, and nucleotide fragments, then this ordered state might look very much like a living organism. Living things are very definitely open thermodynamic systems, existing far from equilibrium. Life uses the energy sources in the environment to maintain an ordered state in apparent defiance of the second law. We've already seen that no real violation of the second law occurs, however, because living organisms also reject their excess entropy to the environment. The details of how existing organisms do this are still being studied by biologists and biochemists, but the underlying idea is not mysterious. The mystery has been this: how did this process begin before the organisms existed? In other words, how did life begin?

The emergence of order in nonequilibrium open systems provides a clue to the solution of this mystery. Accumulating evidence suggests that a spontaneous ordering tendency is quite common in complex systems;

some people (e.g., Stuart Kauffman) believe that such ordering is not only common but almost inevitable. Just as a Bénard cell organizes itself into convection rolls when driven far enough from equilibrium, a complex collection of autocatalytic reactions organizes itself into a stable self-sustaining system, a primitive precursor of life. In this picture, changing conditions will perturb the self-organized ordered state, which may respond by switching to another allowed state more suited to the new environment. This process would be the beginnings of evolution by natural selection, the genetic code serving primarily as a record of the changes. The development of an organism results from the interplay between the information preserved in the genetic code and the natural ordering tendency inherent in the system. These ideas are still novel and speculative. We don't yet understand how life began. The recent work on open systems and self-organization holds great promise, however, to continue providing new insights.

In addition to specific applications, the recognition of emergence is itself of fundamental importance in the sciences. To reiterate, we mean by "emergence" the idea that new properties arise from interactions in complex open systems. These new properties of the system are not predictable, even in principle, from the properties of the individual interactions. Rather, the new properties emerge from the complexity of all the interactions together. This directly contradicts the old reductionist ideal of understanding a system by splitting it into its component pieces (see chapter 14). Here, the behavior of the system is governed not by the pieces but by the system as a whole. Science is still assimilating this new point of view.

§3. ORDER HIDDEN WITHIN DISORDER—NONLINEAR DYNAMICS AND CHAOS THEORY

Unpredictable behavior doesn't require a high degree of complexity to occur. A major new insight in the last few decades has been the realization that even simple systems can act unpredictably. A pendulum, for example, is fairly simple. An ordinary (unforced) pendulum is the epitome of regular predictable motion (that's why we once used them to run clocks). But imagine a pendulum that is given a push at regular intervals (called a forced, or driven, pendulum). This forced pendulum, if it is pushed with enough force at the appropriate frequencies, can go into chaotic motion; the swinging of the pendulum is irregular, and we can't predict what it will do next. Though it's unpredictable, this motion is not random. We'll explore the difference between unpredictable and random as we go along.

For now, my main point is this: Even a simple thing like a pendulum can exhibit complex behavior. Why should this happen? What properties of a simple system might cause chaotic motion to occur? We'll answer these questions as we go along, too. Next, however, let's try to better understand what chaos is by looking at a particular example.

The Logistic Equation

Let's look in more detail at another case that is deterministic, simple, and yet still chaotic. This case, sometimes called the logistic equation, is an elementary algebraic relation

$$x_{n+1} = rx_n(1 - x_n),$$

in which we start with some specific number for x_0, the initial value of x. This initial value is, as you can see, the $n=0$ case. The constant r in this equation is just a number we can choose, sometimes called a parameter (as opposed to the x's, which are variables). Once we have chosen x_0 and r, all the rest of the x_n values are automatically predetermined; that's why we call this a deterministic equation. The way it works is this: Insert the initial value x_0 ($n=0$) into the right-hand side of the equation, and you produce x_1 ($n=1$) on the left-hand side. Take this x_1 ($n=1$) value, and now insert it into the right-hand side. Now you get x_2 ($n=2$) on the left-hand side. Next insert x_2 in order to get x_3, and so on up to any n you please. The appearance of the equation is perhaps deceptively simple. There is actually an iterated feedback loop (see chapters 6 and 21) implied by this method. The behavior of this system (that is, the sequence of x_n numbers we generate) can exhibit an astoundingly rich variety, depending on the values of the parameter r and of x_0.

If we restrict the initial value x_0 to numbers less than 1 and don't let r be any greater than 4, then we guarantee that x_n is never outside the range of 0 to 1. In Figure 12, we graph successive values of x_n against n for the case when $r=3.4$ is the parameter and $x_0=0.02$ is the initial value (n from 0 to 25). You can see that the system soon settles down to alternating between just two x_n numbers, 0.45 and 0.84. We might call this behavior periodic, analogous to the periodic swinging of an ordinary pendulum. The value of r decides the behavior of this system. x_n might settle down to a single repeating number, or might alternate between two (as we saw), four, eight, and so on numbers, depending on r. In Figure 13, we graph the results for $r=4$ and $x_0=0.8$ (n from 0 to 40). The behavior of x_n seen in Figure 13 is quite remarkable; there is no repeating pattern at all. The system now exhibits chaotic variation with n. Because there is no regular-

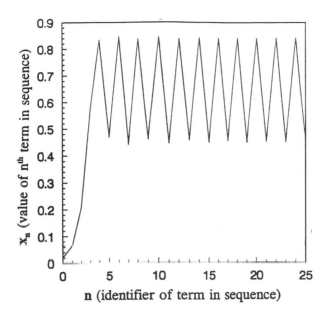

Figure 12. Graph of the logistic equation for $r = 3.4$, illustrating periodic behavior.

ity, we can't predict what the graph will look like for further increases of n (in contrast to Figure 12). The changes are not random, however. We can certainly calculate the next set of x_n values without any ambiguity; but we can't say what they'll look like before we calculate them. In essence, this is what we mean by deterministic chaos.

Sensitivity to Initial Conditions

Chaotic systems have many interesting properties, one of which is illustrated in Figure 14. This graph shows two sets of x_n values (n from 0 to 18). One of these is just the same as the graph shown in Figure 13, except for the shorter range of n (in other words, the dotted line graph repeats the results for $r=4$ and $x_0=0.8$). The second graph is for the same parameter r and starts with an initial value x_0 that is different by only one part in a thousand (i.e., $r=4$ and $x_0=0.8008$). The striking feature that you notice in this graph is that the two results are completely different after about $n=10$. The points start out very close, but they soon diverge from each other a little and the divergence rapidly grows. This property, known as sensitivity to initial conditions, is shared by all chaotic systems.

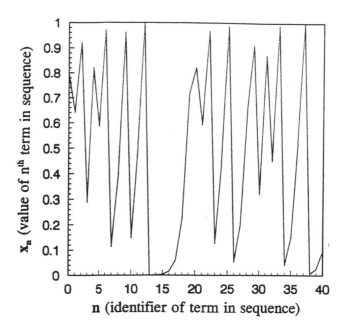

Figure 13. Graph of the logistic equation for $r = 4.0$, illustrating chaotic behavior.

Sensitivity to initial conditions is a very important attribute of chaotic systems in nature. Real quantities in natural systems always suffer from microscopic fluctuations. Even though a chaotic system is deterministic, the long-term behavior of the system is truly unpredictable due to sensitive dependence on initial conditions. The meteorologist Edward Lorenz has called this the butterfly effect; a butterfly flapping its wings in Argentina might start a chain of events leading to a hurricane in Cuba. Lorenz discovered sensitivity to initial conditions while solving a set of model equations describing the atmosphere (the model, although highly simplified, turned out to have chaotic solutions).

Nonlinearity

Why are some systems well-behaved and orderly, while others are irregular and chaotic? The most important characteristic of a system in this regard is undoubtedly whether the system is linear or nonlinear. Chaotic motion is virtually always due to some nonlinearity in the dynamics. In a linear system, the output is directly proportional to the input (we'll look in more detail at linear variation in chapter 19). In contrast, the output

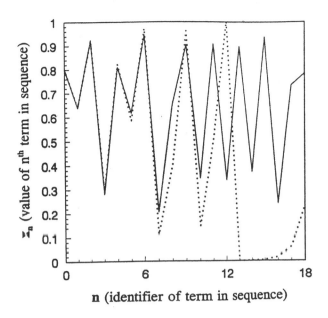

Figure 14. Graphs of the logistic equation for $r = 4.0$, with initial values of 0.8 and 0.8008; the diverging behavior of the two graphs as n increases illustrates sensitivity to initial conditions.

of a nonlinear system depends in some more complicated fashion on the variables of the system; the dependence is not simply a direct proportionality. For example, in the logistic equation that we have explored, the output (x_{n+1}) has a quadratic dependence on the input (x_n), because $x_n(1-x_n)=x_n - x_n^2$. In the driven pendulum example, the effect of gravity on the bob has a trigonometric dependence on the angle (a $\sin(\theta)$ variation), which is nonlinear. In other cases, we might have the product of a variable with its own rate of change; in systems with more than one variable, the output might depend on the product of these variables.

The typical behavior of a linear system is stable and predictable, often periodic. The linear equations governing the dynamics can usually be solved, so we can predict the motion in practice as well as in principle. Also, many systems that are actually nonlinear can be well approximated by linear models (see chapters 6 and 19). The pendulum falls into this category if the arc it swings through is not too large (that's how we get regular motion). Another famous example is the solar system, where the governing equations can be made linear by including only the influence of the sun on each planet, leaving out the effect of all the other planets.

These linear systems, being both important and solvable, received almost all of the attention of scientists for many years. More recently, the widespread occurrence of nonlinearity in nature has drawn greater attention. As Lorenz emphasized, the dynamics of the atmosphere are nonlinear. Turbulence in fluids is governed by nonlinear equations, ecosystems are not linear systems, and many physiological processes are nonlinear. As we have seen, such nonlinear systems are prone to chaotic and unpredictable behavior.

Phase Diagrams

There is an important difference between unpredictable changes and random changes. The difference is subtle and has only been studied in recent decades. The standard meaning of chaos in the English language implies both randomness and unpredictability, a kind of pure disorder. The word "chaos" has a somewhat different technical meaning in the science of nonlinear systems: deterministic yet unpredictable, unpatterned yet non-random. In this sense, the choice of "chaotic" as a label for these systems is confusing, although it has the virtue of being evocative.

In order to illustrate the difference between chaos and random motion, we need to make a brief digression: an introduction to phase diagrams. To understand a phase diagram, we can start by looking at the kind of information we're interested in. We have so far been talking about the form of the time variation in a dynamical system (periodic or irregular, for example). By graphing the important variables of the system versus time, we can see all at once what the time behavior of the system is. The graphs of Figures 12, 13, and 14 are examples of this kind of information, since the successive n values can also be thought of as time intervals (greater n means later time). But if a system has at least two important variables that change with time, we can look at a different kind of information; we can look at how the variables relate to each other instead of how each individually varies with time. In particular, we can make a graph of one variable versus the other variable. For each instant of time, we have a point on the graph representing the values of the two variables at that instant. A long stretch of time will produce many such points, and all the points together make up what we call a phase diagram. If the variation is periodic and regular, the points will form a closed curve with some characteristic shape. A nice example of a simple phase diagram, for an ordinary pendulum, is given in Figure 15. The position of the pendulum bob is plotted on the x-axis, and the velocity of the bob is plotted on the y-axis. As the pendulum goes back and forth, its location on the phase diagram moves from point to point around the curve. One complete pe-

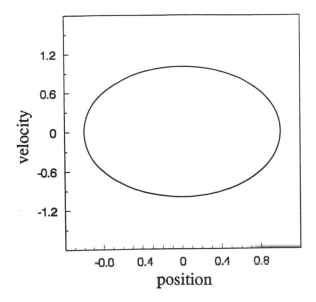

Figure 15. Phase diagram for a simple pendulum, showing the relationship between position and velocity for this periodic system.

riod of the repeated motion represents one complete circuit around the curve. As the pendulum repeats its periodic motion over and over, its point on the graph continues to cycle around.

Strange Attractors

If the behavior of a variable is truly random, then the location of its point on a phase diagram will also change randomly as time goes on. For a system with two randomly changing variables, then, we would expect to see a random distribution of points on the x-y plane of a phase diagram (like raindrops on a car windshield). As we said before, a forced pendulum can go into chaotic motion. The position and the velocity of the chaotic pendulum are both irregular and unpredictable as time goes on. But they are not random. Figure 16 shows a phase diagram for a chaotic pendulum, and this array of points is clearly not random. The motion, though unpredictable, always winds up having a location in phase space that is somewhere on the complicated pattern shown in the figure. This complicated pattern is an example of what has come to be known as a "strange attractor." A strange attractor is a region of a phase diagram on which

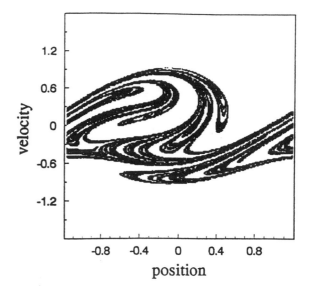

Figure 16. Phase diagram for a chaotic pendulum. The pattern shown, representing the motion of this aperiodic system, is called a strange attractor (image courtesy of J.C. Sprott)

points representing the behavior of a chaotic system are located. Put differently, the motion of the system is attracted to some region of phase space, and we call this region a strange attractor. The word "strange" is used to distinguish these chaotic cases from periodic behavior, where the closed curve is also an attractor in phase space. The strange attractors are very different, though, because there is no simple way to relate the strange attractor on the phase diagram to the observed behavior of the system, which gives every appearance of being random. The existence of strange attractors in chaotic systems is telling us something very profound about nature: *Within the disorder of the chaotic system, there is to be found a hidden order of a different kind.* Strange attractors have many extraordinary properties. They never close in on themselves (i.e., never repeat); instead they fold, twist, and convolute in ever-increasing density and complexity. Points that start out arbitrarily close to each other on the attractor become widely separated as time goes on (this is another result of sensitivity to initial conditions). Because of the never-ending twisting and folding, strange attractors cannot be identified as ordinary lines or planes. Instead, they have what mathematicians call a fractal geometry (fractals are quite interesting in their own right). And of course, we can't fail to notice the mysterious beauty of the visual patterns that strange attractors present.

Chaotic Systems in Nature

All of this fascinating mathematics would be only slightly useful in science if these chaotic systems were an occasional curiosity in nature. Instead, however, we find chaos in a wide variety of natural systems. Regular and predictable behavior seems to be the exception, not the rule. We overlooked this chaotic motion for hundreds of years by concentrating on the problems that we could solve, namely, predictable systems. Irregular behavior seemed uninteresting; we dismissed these as cases where we knew the procedures to find a solution, but actually finding it was too hard. We are now beginning to understand the rich structure of these apparently random cases, and we are discovering how frequently they occur in nature. Since chaotic systems are found in biology, physics, chemistry, and geology (and also the social sciences), and all of these systems share similar characteristics, nonlinear dynamics is also an integrative and unifying idea in the sciences, tying together many otherwise unrelated areas of study.

Many examples of chaos are simple everyday occurrences. The smoke rising up from a lit cigarette or incense stick forms a set of swirls that curl and break up in a chaotic pattern. The creation of a snowflake is a delicate balance between the regular formation of a crystal lattice and the nonlinear growth at its edges (this accounts for the old adage that no two snowflakes are alike). Turbulence in fluids is another example of chaos. The pattern of whirlpools and eddies in a stream is a familiar example of turbulent flow. Water in pipes and the movement of air across an airplane wing also exhibit turbulence under some circumstances, and these are major engineering problems. Turbulence, in fact, was one of the areas in which chaos theory first became prominent. A number of electrical circuits have nonlinear circuit elements, and these circuits produce voltages that vary chaotically. Several such circuits (e.g., the van der Pol oscillator) have been studied extensively, including the properties of their strange attractors. Meteorology is another field in which chaos made an early appearance, when Lorenz discovered that his computer model of the atmosphere produced no predictable pattern (the implications for our ability to make long-term weather predictions are apparent). In the course of his studies, Lorenz found a strange attractor for his system, shown in Figure 17, which has become quite famous.

In biology, an example of chaotic dynamics is found in the study of population changes. In fact, the logistic equation we worked with is a simple model of how populations change from year to year. (The variable x, ranging between 0 and 1, represents the population divided by its maximum value.) In this model, n means the number of years, x_0 is the initial population, x_n is the population in year n, and r is a measure of the fertility

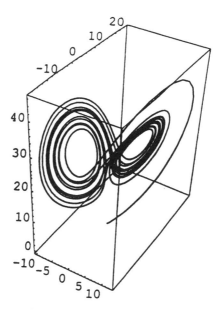

Figure 17. The Lorenz attractor, showing the relationship between three variables in a chaotic climate model.

of the species. The population in year $n+1$ is proportional to the population x_n in year n (for obvious reasons), but also proportional to $(1 - x_n)$ because resources become more scarce as the population grows. For moderate values of r, the population settles to a stable number, or cycles regularly between a few numbers. But as we have seen, populations can also vary chaotically from year to year. Even this simple model captures important features of real changes in some cases (e.g., certain species of insects).

Perhaps the most dramatic example so far is the discovery of nonlinear dynamics and chaos in the workings of the human heart. We think of the heartbeat as a regular and periodic occurrence, but the heart needs to change the tempo of its rhythmic pumping over a wide range while still remaining stable. The operation of the heart is extremely complex, with the muscles and nerve stimuli of many different parts all working together at the proper times. A linear oscillator, if it's knocked out of normal operation, has a difficult time regaining this normal operation. A little nonlinearity mixed properly into a system actually can help the system maintain a robust stability (which we certainly want in the operation of our heart). As we've seen, however, small changes in a system parameter can drive the system from stable behavior into the chaotic regime; in the heart, this means irregular fibrillations and possible death. So order and disorder coexist in the heart, and both must play their roles properly for the heart to work.

FOR FURTHER READING

Order and Chaos, by S. W. Angrist and L. G. Hepler, Basic Books, 1967.
Engines, Energy, and Entropy, by John B. Fenn, W. H. Freeman, 1982.
Chaos, by James Gleick, Viking Penguin, 1987.
Deterministic Chaos, by H. G. Schuster, VCH, 1989.
Exploring Complexity, by G. Nicolis and Ilya Prigogine, W. H. Freeman, 1989.
Applied Chaos Theory, by A. B. Cambel, Academic Press, 1993.
Complexification, by John L. Casti, HarperCollins, 1994.
What Is Life? The Next Fifty Years, edited by M. P. Murphy and L. A. J. O'Neill, Cambridge University Press, 1995.
Website: http://sprott.physics.wisc.edu/fractals/

EPIGRAPH REFERENCES: Lao Tsu, *Tao Te Ching* (translated by Gia-fu Feng and Jane English), Vintage Books, 1972, verse 45. H. Margenau, quoted in "The Question of Reform," Project Kaleidoscope, 1997, p. 17. G. Nicolis and I. Prigogine, *Exploring Complexity*, W. H. Freeman, 1989, p. 8. Heraclitus, fragments, quoted in *The Origins of Scientific Thought*, by Giorgio de Santillana, New American Library, 1961, p. 44.

Chapter 18

RIDING BLAKE'S TIGER: SYMMETRY IN
SCIENCE, ART, AND MATHEMATICS

> But the idea of the crystal is nothing but its spatial
> symmetry. . . . all the actual forms of crystalline
> symmetry, and only these, follow from the mathematical
> characteristics of Euclidean space.
> *(C .F. von Weizsacker)*

> Symmetry is a vast subject, significant in art and nature.
> *(Hermann Weyl)*

THE CONCEPT of symmetry plays an important role in all of the natural sciences, playing a particularly fundamental role in physics. This concept is also of prime importance in mathematics, where our intuitive notions of symmetry gain precision and rigor. Symmetry is central to aesthetics and the arts, but also useful in technical work and engineering. Few concepts have such wide-ranging implications. What does the word "symmetry" mean exactly? In general usage, symmetry implies a sense of being harmonious and well balanced. In geometry, it has a more restricted meaning: If you do something to a geometric figure (move it, spin it, flip it over, etc.), and it still looks the same, then we say that the figure has a symmetry. These two definitions are not as far apart as they may at first seem. We'll begin by exploring the geometric concept of symmetry, and then extend our understanding of its applications in science as we go along. In the process, we'll make contact with the general aesthetic sense of symmetry; we'll show how it serves as a unifying principle in science; and we'll discover its deep ramifications in mathematical and physical theory.

§1. SOME BASIC IDEAS

Rotations, Reflections, and Translations

Take a look at the cross with four equal arms shown in Figure 18a. Suppose that I rotate this cross (about its center) through one-quarter of a full circle. We are left with a cross that looks identical; we say that this cross is symmetric under the operation of one-quarter turn rotations. We

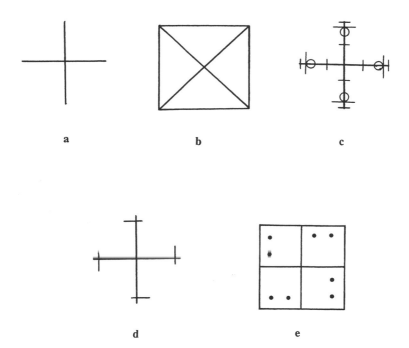

Figure 18. Some shapes with fourfold rotational symmetry. Three also have reflection symmetry, but two do not.

could equally well have done a three-quarter turn, a half turn, or a full turn rotation. For any of these operations, we get a cross that looks identical to the original (for any other angle, we get something different). We say that this cross has a four-fold rotational symmetry. For all of the other shapes in Figure 18, the angles that bring us to an unchanged shape are the same angles we found for the cross, namely, integral multiples of quarter-turns. All of these shapes have a four-fold rotational symmetry. You can see intuitively that they share some property, but now we have precisely defined what that property is. Any conclusions that apply to fourfold symmetric objects now apply automatically to each of these shapes and to all other imaginable shapes sharing this symmetry.

There is a subtle difference between the first three shapes (a, b, c) and the second two (d, e). This difference, though subtle, is quite important (even profound). To see the difference, imagine a vertical line (in the plane of the paper) through the center of each shape, and then fold over this line as if you were folding the paper over. You see that three of the shapes have matching halves under this operation, while the other two do not. We say that the first three have a bilateral symmetry (or reflection symme-

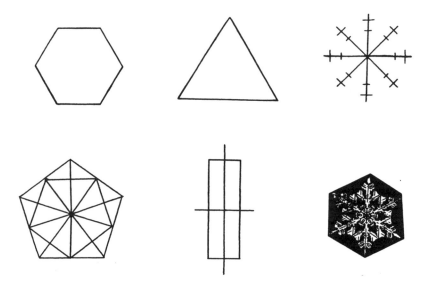

Figure 19. A variety of shapes with *n*-fold rotational symmetries.

try), while the second two don't. This kind of symmetry is familiar; for example, anyone who has cut out paper dolls has created bilateral symmetry. If you imagine placing a mirror on the imaginary line through the center of the shape, you'll also create a bilaterally symmetric figure. This is the source of the term *reflection* symmetry. A shape can have a number of reflection symmetries along different lines (vertical and horizontal in these particular cases). Although our examples so far have had both rotational symmetry and bilateral symmetry, a shape can easily have a bilateral symmetry without a rotational symmetry (a horseshoe, for example). We can have shapes with any number of rotational symmetries (*n*-fold symmetry, where *n* is an integer). Some examples are given in Figure 19 (they also have some reflection symmetries). Note that all of the so-called regular polygons have *n*-fold rotational symmetries. The ultimate in rotational symmetry is the circle: it can be rotated through any angle at all without being changed. Any line through the center of a circle results in a bilateral symmetry.

The other major type of symmetry is called translational symmetry. Imagine that each point in a shape is moved over a given distance, in a given direction. The result of this process will be a reproduction of the shape. Now repeat the operation, over and over again forever. This is translational symmetry. "Translate" simply means "move to a new spot." If we have translational symmetry, then we move to a new spot that looks

Figure 20. A pattern with two-dimensional translational symmetry.

exactly like the old spot. These ideas are consistent with our definition of symmetry given above (if you do something to it, it still looks the same). An interesting example of translational symmetry is shown in Figure 20.

Symmetries in Three Dimensions

So far, all of our work has been in two dimensions, that is, in a plane, because it's much easier to visualize (and draw) figures in a plane. Most of the important ideas can be introduced and illustrated without the complications of three dimensions. When we go from two dimensions to three dimensions, however, some new and interesting features appear. In two dimensions, we rotated about a point in space, but in three dimensions we must rotate about a line (called the rotation axis). Because this line must have a direction, and it might be any direction, this case is rather more complicated. The symmetry properties of the shape will be very different for different choices of the rotation axis direction. The only exception to this statement is the sphere. For any line through the center of a sphere (any line at all in any direction) the sphere can be rotated through any angle and remain unchanged. The sphere has the same high degree

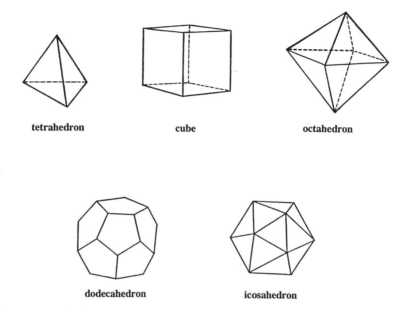

tetrahedron   **cube**   **octahedron**

dodecahedron   **icosahedron**

Figure 21. The five Platonic solids. These are the only possible shapes that have identical regular polygons for every side.

of rotational symmetry as the circle, and has it for any rotation axis (through its center) that we choose. (Any plane through the center of the sphere is also a plane of reflection symmetry.) Perhaps this remarkable degree of symmetry is the reason the Greeks considered circles and spheres to be perfect figures (see chapter 5). Another fascinating discovery about rotational symmetry in three dimensions goes back to the Greeks, this one a bit more complicated. In a plane, a regular polygon has n equal sides, n equal angles, and n-fold rotational symmetry. Starting with $n=3$, we see that n can be any number we please. Regular polygons might be hard to draw for some values of n, but they aren't hard to imagine. The situation is quite different in three dimensions. The three dimensional analog of a polygon is a polyhedron, a solid figure with polygonal faces (for example, a cube is a polyhedron having six square faces). A *regular* polyhedron has only identical regular polygons for each of its faces (so a cube is also a regular polyhedron). While there are an infinite number of regular polygons in two dimensions, it turns out that there are only five regular polyhedra in three dimensions (they are shown in Figure 21): the tetrahedron with four triangular faces; the cube with six square faces; the octahedron with eight triangular faces; the dodecahedron with twelve

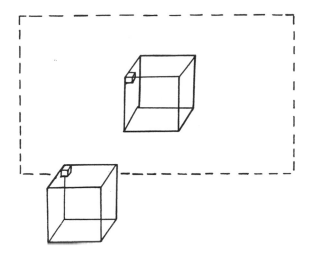

Figure 22. A reflection symmetry in three dimensions. It's impossible to make the two shapes identical by any combination of rotations and translations.

pentagonal faces; and the icosahedron with twenty triangular faces. These shapes are known as the Platonic solids, and they have an intriguing array of symmetries.

Reflections in three dimensions also present some novel features. In Figure 22, we see two cubes, each having a marked corner. These cubes have a reflection symmetry in the indicated plane. This plane is sometimes called a mirror plane, for obvious reasons. Now, try to imagine moving these two cubes around so that one of them is perfectly superimposed on the other. You can't accomplish this task. In three dimensions, no set of rotations and translations is equivalent to a reflection. The same thing happens if you look at yourself in a mirror and try to move your right and left hands around so that they match—it can't be done. For translations in three dimensions, we need to know several directions and distances in three dimensions to define the translational symmetry. Translationally symmetric figures in three dimensions are more difficult to visualize, but don't present many new conceptual features. Translations combined with other operations are more interesting. For example, we can combine a translation in one direction with a rotation around that direction, resulting in a helical shape, also called a screw. The mirror reflection of this helical screw is another helical screw, and these two shapes, though symmetric under reflection, are intrinsically different from each other. One is called a right-handed helix, and the other is called a left-handed helix (a right-handed screw goes inward when it is turned clockwise, while a left-handed screw comes out when it is turned clockwise). As we'll see later, these possibilities are realized in nature with fascinating consequences.

§2. USEFUL AND BEAUTIFUL APPLICATIONS OF SYMMETRY

Chemical Bonds

Now that we know something about symmetry, let's explore how it relates to science. We'll start with the simple methane molecule (methane is the main constituent of natural gas), which consists of one carbon atom joined to four hydrogen atoms. The carbon atom is at the center of a regular tetrahedron, one of the Platonic solids (see Figure 21). The hydrogens are located at the corners of this tetrahedron. Tetrahedral bonding is common, in part because this is the only way in which four atoms can be bonded together in a three-dimensionally symmetric fashion. The electrons of the atoms making up the molecule also have their own symmetries, which partly determine the symmetry of the bonds they make. Metal atoms typically have different symmetries than carbon, and metal atom sites often play a key role in biologically important processes. Hemoglobin, for example, is the component of your red blood cells that carries the oxygen to the rest of your body. The iron atom in the substance called heme is bonded to four nitrogens at the corners of a square, with the iron at the center (see Figure 23). The heme is bonded to a large protein molecule called globin at a single bonding site: the iron atom. This iron atom also forms a bond with a water molecule. So iron is bonded to six other atoms, and these bonds form an octahedral symmetry (another Platonic solid; see Figures 21 and 23). Four bonds are to the nitrogens in the heme, one bond is to the globin, and the last bond is to the water molecule, while the iron atom sits at the center of the octahedron. The water is replaced by oxygen in the lungs to make oxyhemoglobin, and then the oxygen is replaced again by water when the oxygen is delivered to the cells of the body. A similar octahedral bonding structure is found in chlorophyll, the substance that plants use to photosynthesize food. In the case of chlorophyll, the metal atom found in the octahedrally symmetric bonding site is magnesium.

The importance of bonding, structure, and symmetry is dramatically illustrated by the classic case of pure elemental carbon. Carbon atoms can bond to each other in several ways. They might form tetrahedral bonds (like methane), but with carbon at the center and also at each vertex of

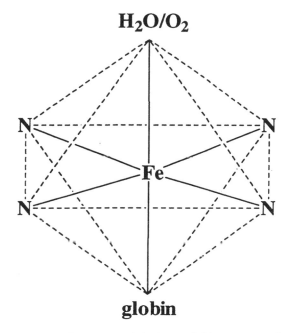

H₂O/O₂

globin

Figure 23. A schematic illustration of the hemoglobin structure, showing the octahedral symmetry of the iron atom binding site. The alternate binding of water and oxygen molecules to the iron is the means by which blood transports oxygen from our lungs to our cells.

the tetrahedron. With each carbon atom bonding to four other carbon atoms, every carbon atom is tetrahedrally bonded to four more in an endless interlocking array. Carbon in this structure is what we call diamond. But, carbon can also form bonds to three other carbon atoms within a plane, forming flat hexagonal arrays that are only weakly bonded to each other. We call this form of carbon graphite, the stuff your pencil leads are made of. The properties of graphite and diamond are very different, yet both are made only of carbon atoms. Just a few years ago, chemists created a new form of carbon in which sixty carbon atoms are bonded together into a soccer-ball-like structure that was named buckminsterfullerene. This structure of interlocking hexagons and pentagons also has a variety of interesting symmetries. As we've seen, symmetry plays a central role in structure and bonding. In the examples of graphite and diamond, we also see some new features not found in molecules. The arrays of atoms go on repeating forever, so we now have something new: translational symmetry. We are now talking about crystals.

Crystallography

Crystals have captured the imagination of humans since before recorded history. Small crystals have been found among the power objects of shamans at archaeological sites, the ancient Egyptians venerated the mineral lapis lazuli, and almost every great monarch has hoarded gemstones. The beauty of gems and crystalline minerals is partly in their colors and brilliance, but it's also partly in their highly facetted geometric shapes. Today, crystals are valued not only for their beauty but also for their technological usefulness. As one example, the microelectronic integrated circuits used in computers, stereos, televisions, and other electronic devices are manufactured from single crystals of the element silicon. Inherent in the beautiful geometric shapes of crystals are many symmetries, including examples of each type (rotational, translational, reflection) we've discussed. Ice crystals, for example, have a six-fold rotational symmetry, which is revealed in the spectacular beauty of snowflakes. The science that studies the structures of crystals is known as crystallography, and we aren't surprised to find the use of symmetry as one of its central organizing principles. Crystallography in turn is foundational to the physics and chemistry of solids, and also to mineralogy and the earth sciences. An interesting historical note: The mathematical basis of crystallography, which is just the systematic investigation of all the possible symmetries that a crystal might have, was completed by the end of the nineteenth century, before the experimental confirmation of the idea that crystals are regular and periodic arrays of atoms. The defining characteristic of crystals is their translational symmetry. Crystals are ordered periodic arrays of atoms, which repeat regularly over and over again in space. The local structure of the crystal is then reproduced at any equivalent point, and the overall (macroscopic) structure of the crystal inherits any symmetries that the local structure might possess. This repeating local structure in a crystal is called its unit cell. If the unit cell has a four-fold rotational symmetry, for example, so does the entire crystal. In this way, the combination of local symmetry in the unit cell along with translational symmetry gives rise to the final symmetry of the crystal structure. This symmetry, in turn, manifests itself in the facets and angles of the actual crystal that we see. The beautiful shapes of crystals are ultimately due to the ordered arrangements of their atoms.

A simple real-life example of a crystal known to all of us is table salt, sodium chloride. A careful look reveals that salt crystals are cubic. The atomic structure of salt, shown in Figure 24, is actually two interlocking cubes (one of sodium and one of chlorine), so each kind of atom is surrounded by neighbors of the other kind. Because each crystal face consists

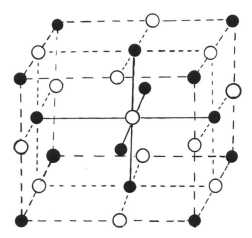

Figure 24. The crystal structure of sodium chloride (table salt). Filled circles represent the sodium ions; open circles represent the chlorine ions.

of squares, with atoms at the corners and one atom in the center, this type of structure is known as a face-centered cubic crystal (this is one of the most common structures for pure elements). This crystal has three perpendicular rotation axes (one for each dimension), each of which has four-fold rotational symmetry. There are other rotation axes in the cubic salt structure that have three-fold symmetry and two-fold symmetry. Each rotation axis is perpendicular to a plane of atoms, and each of these planes is a plane of reflection symmetry.

The need for translational symmetry in three dimensions restricts the kinds of rotation and reflection symmetries that a crystal might have. For example, the only rotational symmetries that are allowed are 2-fold, 3-fold, 4-fold, and 6-fold. Any other type of rotational symmetry leads to a structure that can't be repeated translationally (and still fill all space without any gaps). This restriction accounts for the fact that no crystal can have pentagonal symmetry (although a fascinating modern development has been the discovery of quasicrystals, which have 5-fold rotational symmetry and almost, but not quite, have long range order). Crystallographers have worked out all of the possible combinations of rotational, translational, and reflection symmetry that can exist in a crystal; a total of 32 symmetry classes turn out to be possible. The 32 symmetry classes can be further categorized into 6 sets that have various symmetries in common. These 6 sets are the major crystal systems used to categorize minerals and other crystalline solids (see Figure 25). The different crystal systems are characterized by the angles between their axes and by the relative lengths of these axes. In the isometric system, for example, all the

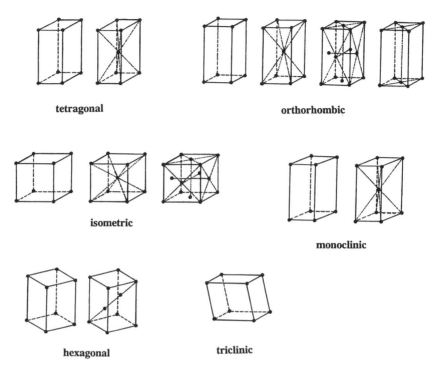

Figure 25. The fourteen Brevais lattices, organized into the six major crystal systems. Crystals are limited to these possible structures by fundamental symmetry considerations.

lengths are equal and all the angles are right angles. In contrast, the triclinic system has no right angles and no side length equals any other. Between these two extremes, we have the orthorhombic (all right angles, no equal sides), tetragonal (all right angles, two equal sides), hexagonal (two right angles, two equal sides), and monoclinic (two right angles, no equal sides) systems.

The possible types of "space lattice" structure are also limited. In addition to the face-centered structure we've seen in salt, the unit cell might also be body-centered, side-centered, or primitive. Combining these possibilities with the 6 crystal systems results in the 14 so-called Brevais lattices (Figure 25). Any crystal structure, no matter how complicated, can be ultimately reduced to one of these 14 potential structures. A number of minerals and ceramics, with complicated compositions and a variety of bonding characteristics, exhibit highly complex crystal structures (each point on the lattice, which is the translationally repeating unit, can have a group of atoms with its own structure). Even the most complicated crystal structure, however, must obey the constraints imposed by fundamental symmetries.

As we've seen in the examples of diamond and graphite, the structure (and symmetry) of a crystal has a profound impact on its physical properties. Physical properties can also affect symmetry quite directly. The open crystal structures of silicon, germanium, and diamond, with their tetrahedral symmetry, result from the strongly directional covalent bonding of their constituent atoms. Sometimes, the symmetry properties of a crystal affect its properties in a direct and unambiguous manner. For example, the piezoelectric effect (producing electricity by squeezing a crystal) cannot occur in a crystal that has a reflection symmetry plane perpendicular to the direction you squeeze the crystal. Optical properties, such as the way a crystal polarizes light, are directly affected by crystal symmetry. A dramatic example is double refraction by calcite crystals, in which an object seen through the crystal is split into two separate images. Symmetry alone doesn't determine physical properties (metals as dissimilar as sodium and iron both have the same body-centered cubic structure). On the other hand, only the translational symmetry of crystals has enabled scientists to understand the properties of materials, with their uncountable numbers of electrons. Only the fact that all of the electrons share a similar environment in all parts of the sample (in other words, translational periodicity) makes this problem tractable. Taking advantage of this symmetry, we are able to understand a great deal about the properties of metals, insulators, and semiconductors (see chapter 1).

Points, Spheres, and Tires

To end the section, let's treat two new topics that are completely different. One of the simplest symmetries imaginable in three dimensions is that of a single point. A point looks the same from any angle that you look at it. Putting this in fancier language, we say that a point has full rotational symmetry about any axis (through the point). We've seen this symmetry before: the rotational symmetry of a sphere (with our point at its center). Spherical symmetry is intimately related to the many inverse squared distance relationships found in science (examples are Newton's law of gravitation, Coulomb's law for electrical charges, and the decrease of light and sound intensity with distance). The point source (of mass or charge, for example) can be replaced by any spherically symmetric distribution without changing any effects outside the sphere. These results follow from combining spherical symmetry with the relevant physical laws in each case.

We'll now leave such highly abstract matters to consider a practical problem faced by most of us on occasion: balancing automobile tires.

What does it mean when your mechanic asks you if you want your tires balanced? Ideally, a tire should have full circular rotational symmetry about its rotation axis (which is the axle of the car). If it doesn't, then the tire will have a pronounced tendency to wobble at high rotational speeds. This wobble is felt by the driver as an unpleasant vibration of the whole car. Now, due to small manufacturing inaccuracies, the tire's axis of symmetry and rotation axis may not coincide exactly. By attaching small weights at the right places on the edge of the wheel, the mechanic brings these two axes into alignment. The rotational symmetry of the tire is restored. This is called balancing the tire.

§3. Some More Ideas

We have so far discussed cases in which the symmetry is perfect, though we've already had to admit this is not always realized in nature. In this section, we'll explore more fully the concepts of approximate symmetry and broken symmetry. In addition, we'll introduce a somewhat more abstract notion of symmetry, not so tied down to geometry. This is the mathematical concept of the transformation. Our thinking about transformations will lead naturally to a brief discussion of the mathematical concept of a group, and the relationship between symmetry and group theory.

Broken Symmetry

To think about broken symmetry, consider this example: If we take a sphere and drill a hole in it, then we will obviously lose the spherical symmetry. But, we'll still have a figure with circular symmetry (about an axis through the center of the hole and the center of the sphere). This example is typical. A broken symmetry is the introduction of some lower-symmetry element into a highly symmetric figure. In nature, it's not uncommon for such symmetry breaking to occur in a seemingly symmetric system, sometimes due to small random fluctuations. Imagine a straight line of ants coming to the center of a barrier in their path. The symmetry of the system suggests that the ant line will split up into two halves going around the barrier, but instead the symmetry is broken and they mostly go only one way around. As another example, recall our translationally symmetric crystals and imagine that a single atom is missing (this happens in real crystals, and is called a defect). The symmetry of the lattice has been broken by the defect. Mathematically, the translational symmetry is gone. But for most of the crystal, it's still there; we still have approximate translational symmetry. So for many properties of the crystal, we can ignore the presence of the defect. The broken symmetry is strong in the neighborhood of the defect, though, and local properties are dominated by it. At the edge of the crystal where the lattice ends (the surface), the translational symmetry is broken strongly. Once again, some properties of the crystal are dominated by the presence of the surface, while other properties are unaffected. A person's face is an interesting case of approximate bilateral symmetry. Perhaps you have heard that a slight asymmetry is added to patterns by Islamic rugmakers as an intentional imperfection. Approximate bilateral symmetry is commonly found both in art and in biology.

Transformations and Group Theory

A transformation in mathematics is what the normal usage of the word suggests: a change from one thing to another. The difference from normal usage is that in mathematics, the change is very precisely defined by some rule. For example, we might have a transformation that turns numbers into their doubles (1 into 2, 7 into 14, etc.). The rule in this case is "double this." In the geometric cases we've looked at, the transformations have been movements in space, with rules such as "rotate one-fifth of a turn" or "translate one lattice spacing to the right." The advantage of casting the discussion in terms of transformations instead of motions in space is that we can now broaden the concept of symmetry. We can include new cases, which are not geometric but which follow similar rules. For example, suppose we have a rule that is "turn this number into its negative." This rule is actually identical to a reflection in one dimension about a point (we can associate this point with the number zero). We have found a symmetry in numbers. This case is fairly simple and easy to visualize, but the same kind of thinking can be applied to more complicated cases.

If transformations have some symmetry, then we can combine them in a way that brings the system (or moves the figure) back to its starting point. If we have a reflection symmetry, for example, then two reflections produce our original figure back again. A figure with three-fold rotational symmetry (like a triangle) rotated through three successive rotations gets us back to where we started; it's as if the triangle has not moved. A transformation that leaves you with no change is an important transformation. Mathematicians call this the identity. If two successive transformations result in the identity, we say that one transformation is the inverse of the other. The inverse of a clockwise rotation, for example, is an equal counterclockwise rotation. The inverse of a translation to the right is an equal translation to the left. We've already seen that the inverse of a reflection is just another reflection.

Now, suppose we have a collection of transformations that includes the identity, and for which every transformation has an inverse. Impose one last condition: If we combine any two of these transformations, the result must be another of the transformations that is in our collection. A collection of transformations that has these properties is called a group. The study of the properties of groups is called group theory, and from these simple beginnings it evolves into an important branch of mathematics. Let's look at a simple example of a group of transformations, in order to get a better sense of what groups are. An equilateral triangle has several transformations that leave it looking unchanged. We can rotate it through a one-third turn, a two-thirds turn, or a full (three-thirds) turn. The full turn brings it back to where it started, so that is the identity. In addition to these three angles, we can also do reflections about the lines that go through the vertices, perpendicular to the opposite sides (i.e., lines that bisect the triangle). Since there are three reflection axes, we have six symmetry transformations, one of which is the identity. Now, if we do any two of these consecutively, we can get the same figure by doing one of the others in our collection. So, these six transformations make up a group. This group is called the symmetry group of the triangle. This group of transformations characterizes precisely the symmetry of the triangle. In fact, it characterizes the symmetry of all figures having three reflection symmetries and a three-fold rotational symmetry. The notion we've developed is very general. The symmetry possessed by any shape is specified by its symmetry group. The symmetry group of a sphere, for example, has an infinite number of transformations (all rotations about any line through its center, plus reflections about any plane through its center). The symmetry group for the icosahedron consists of 120 transformations. And Euclidean space itself is characterized by its symmetry group, consisting of all rotations, reflections, and translations. The abstract methods of group theory can be used to derive many important results. For example, a crystal must have a symmetry described by one of 230 possible space groups, and only these. The greatest power of group theory, though, stems from its ability to describe the symmetries of transformations that are not even geometric, providing a common mathematical language with which to describe symmetries both visualizable and nonvisualizable. We'll come back to this point in §5.

§4. Symmetry in Biology and Art

Biology

Among the most apparent symmetries in our experience is the bilateral symmetry of the human being. Bilateral symmetry is a property we share with most of the vertebrates, insects, and other (but not all) members of the animal kingdom. But this bilateral symmetry, although overwhelmingly present in animals, is only an approximate symmetry. We appear on the outside to be bilaterally symmetric, but our internal organs are asymmetric (the heart on the left side, for example). Also, most people have a dominant hand. Even the appearance of symmetry is only approximate, since there are small blemishes, differences in fingerprints, and so on. Another interesting asymmetry is in the functioning of the brain, where some evidence indicates the right and left halves are partially specialized for different thinking processes (spatial, linguistic, musical, etc.). Other animals exhibit quite a variety of symmetries. The starfish has five-fold rotational symmetry, along with five reflection symmetries. The octopus and the medusa have eight-fold symmetry, while the hydra has six-fold symmetry. Many of the microscopic radiolarians have the symmetries of the Platonic solids. Once again, of course, these symmetries are all approximate to varying degrees. It's probably not surprising that living organisms have a lower degree of symmetry than inorganic forms such as crystals. The more interesting point is that they have as much symmetry as they do.

Another extremely interesting question is involved here. There's a little bit of broken symmetry in the developed animal, but there is a huge amount of broken symmetry in its development from an initial single cell. How does this come about? Although the blastula stage of an embryo appears to be spherically symmetric, we know experimentally that there are inherent asymmetries even at that stage. These asymmetries can be traced back to the orientation of the egg, and become more apparent as the embryo develops. A great deal of knowledge has been acquired from the study of embryo development, but this is still an area in which we don't have a fundamental understanding.

Plants also offer some beautiful instances of symmetry. Flower petals characteristically show a number of different rotational symmetries, with five-fold and six-fold being perhaps the most common. Three-fold, four-fold, eight-fold and even higher symmetries are also seen (usually accompanied by their corresponding reflection symmetries). Trees often have an approximate cylindrical symmetry, while some climbing vines exhibit helical symmetries, both right-handed and left-handed. More complicated versions of helical symmetry are seen in sunflower florets and fir cones. The most prominent helical symmetry in biology is undoubtedly the DNA molecule, with its celebrated double helix structure. The lack of translational symmetry in DNA is an important property, since the aperiodic structure of the molecule encodes the genetic information of an organism. Translational symmetry is understandably seen less often in biology, but there are a few approximate cases. In one dimension, the segmented worms, centipedes, and so forth, show this property in animals, while bamboo stalks are an example from the plant world. A striking and fa-

mous illustration of symmetry in nature is the beehive, which has translational symmetry in two dimensions as well as six-fold rotational symmetry. Since a hexagonal lattice of this sort provides the most efficient
packing in two dimensions, the bees manage to minimize their use of labor
and material in their architecture. The geometric ability of the bees is
legendary, and a good deal has been written about the advanced mathematics of three-dimensional beehive structures.

A number of the molecules found in biological organisms do not have
reflection symmetry. Since the molecule is not identical to its mirror
image, these substances have two forms, a right-handed molecule and a
left-handed molecule. Glucose is a prominent example. The molecules are
almost indistinguishable chemically, but sometimes have different effects
on light shining through them. You might expect the two forms to exist
in equal amounts, and they do for substances created in the laboratory.
Pasteur discovered the existence of such substances when he made tartaric
acid for the first time in 1848, and he obtained equal amounts of the
right-handed and left-handed forms. The right-handed form was already
known, having been found in fermenting grapes (wine); the left-handed
form had never been seen before. Only one of the two forms is produced
by the biological process. In fact, living organisms almost always produce
only one form of such mirror image crystals. Our bodies, for example,
contain only right-handed glucose (but left-handed fructose). No one
knows the reason for this broken symmetry. Another biological example
of handedness is found in the shells of snails. A snail shell is a spiral
helix, which could be right-handed or left-handed. Instead of occurring
in roughly equal numbers, almost all species of snails have right-handed
shells. Again, there is no apparent reason for this. (Remarkably, even one
of the basic interactions of nature, the weak nuclear force, exhibits an
unexpected handedness.)

Art

Let's now take a look at symmetry in art. Symmetry often plays a major
role in artwork because there is a strong aesthetic quality in the consideration of symmetry. Once again, however, it's broken symmetry that is
usually important. In art, perfect symmetry would have a tendency to
become monotonous. Bilateral symmetry has been common in many art-
works, from antiquity to the present. Many paintings have an overall
form of bilateral symmetry, but exhibit broken symmetry in their details.
For example, in paintings of the Last Supper, Christ sits in the line of
symmetry with six apostles on either side. The style of Tibetan painting
in which the Buddha sits at the center and various scenes are positioned

Figure 26   Two patterns used in Moorish ornamental artwork.

symmetrically about the painting (each scene being different) is another example. Architecture provides many instances of buildings that are bilaterally symmetric when viewed from the front (Gothic cathedrals, the White House, the Taj Mahal, etc.). Rotational symmetry is less important in painting, but is often found in mosaic tilings, stained glass windows, and so on. We also find rotational symmetry in the ornamentation of vases and the tops of columns in buildings. Buildings themselves occasionally exhibit rotational symmetry, a celebrated example being the leaning tower of Pisa. Translational and reflection symmetries, combined with rotations, are all found in what H. Weyl refers to as the art of ornament. This category includes ceramic tiles, wallpaper, and cloth prints; among its finest realizations are the remarkable achievements of the Arabs in their glass and mosaic art. A famous example of this art is the Alhambra, a Moorish palace in present-day Spain. Two Moorish patterns, taken from Weyl, are shown in Figure 26. The mathematical analysis of these complex patterns is difficult, and it's a testament to the ingenuity of the artisans that they apparently exploited every symmetry available (17 symmetry groups are now known to be possible in two dimensions; examples of every one have been found in Egyptian ornaments). An interesting technique employed by these artisans was to create a geometric pattern with some high symmetry, and then use color to produce a new and lower symmetry in the actual artwork. Once you become aware of their presence and start to notice them, symmetries are overwhelmingly apparent in art, architecture, and ornamental design.

§5. Deep Applications

We've mostly been discussing symmetries that are visualizable and readily apparent. There are also mathematical symmetries in the sciences that are not as obvious but are very important. Particularly in physics, we find that symmetry is fundamental to the workings of nature. Let's look at some of these cases.

Unification

A wonderful example comes from the theory of electricity and magnetism. Around the middle of the nineteenth century, James Clerk Maxwell pondered the known equations governing electricity and magnetism. He noticed an asymmetry in these equations: a changing magnetic field causes an electric field, but a changing electric field doesn't cause a magnetic field. Maxwell rewrote the equations to make them symmetric (mainly by adding one new term), and he then explored the new equations to find out what implications such a change might have. He discovered a remarkable result. The altered equations predicted waves of electric and magnetic fields that travel with the same speed that light was known (experimentally) to have. The correct interpretation of this result is that light is in fact an electromagnetic wave moving through space like ripples on a pond. Maxwell had made a fundamental discovery about the nature of light, an age-old problem, and unified the sciences of optics and electromagnetism. This episode has entered physics lore as the first of a series of unifications grounded in symmetry. The next major unification was due to Einstein, the theory of relativity. Once again, an asymmetry in the laws of physics was perceived. This time, the asymmetry was between two observers in relative motion. To rid physics of this asymmetry, to make the laws of physics the same for all observers, Einstein proposed revolutionary ideas. He welded together space and time into a single space-time continuum, and discovered the symmetries of this unified space-time (physicists refer to these symmetries as the Lorentz transformations). A consequence of unifying space and time was the further unification of mass and energy into a single fundamental entity.

More recently, a goal in physics has been to unify the so-called four fundamental forces. These four forces, which are gravitation, electromagnetism, and the strong and weak forces in atomic nuclei, are presently thought to be the only interactions in nature; others, such as friction and chemical bonding, can be reduced to these four. The goal is to reduce these four to just one. A step in this direction was accomplished when the electromagnetic and weak interactions were unified, that is, shown to be two manifestations of the same underlying interaction. Why then do they appear to be so different? The unified interaction is described by a symmetry, and a spontaneous symmetry breaking occurs resulting in two separate forces. This symmetry breaking would have occurred early in the evolution of the universe, as it cooled down from the big bang. Although the strong nuclear forces haven't yet been included in a unified theory, there has been some progress in understanding these forces based on the whimsically named quark model. The interactions of the particles called quarks explain a great deal of the behavior of elementary particles in general. These quark interactions can be described by a small number of fundamental symmetry operations. The symmetries are not visualizable, but are instead presented in the language of group theory, similar to the manner we've discussed above.

Another important application of symmetry is to the conservation laws of physics. We have already examined the fundamental importance of energy conservation (see chapter 17). A number of similar fundamental conservation laws exist. Momentum, angular momentum, and electric charge are all conserved quantities in both classical and modern physics. In 1918, the mathematician Emmy Noether showed that every conservation law is due to some underlying symmetry. Conservation of momentum results from the translational symmetry of space, for example; conservation of angular momentum results from the rotational symmetry of space; conservation of energy results from the translational symmetry of time (i.e., the laws of physics don't depend on when they are observed). Electric charge, and several other more abstract quantities, are conserved due to the abstract symmetries mentioned previously.

Permutations, Symmetry, and Algebra

For our next example, we need to introduce one more concept, the permutation group. "Permutation" is just a fancy way to say "rearrangement." For example, 213 is a permutation of the digits 123, and there are six permutations of 123 possible. What does this have to do with symmetry? Place the numbers 1, 2, and 3 at the vertices of an equilateral triangle, and perform all possible symmetric rotations and reflections of this triangle. The remarkable result of these operations is that we get the same six permutations! The permutation of three objects has the same symmetry as the triangle, and for that reason is described by the same group. For more than three objects, we no longer have a simple geometric interpretation, but we can still create a symmetry group.

Now, one important postulate of the modern theory of matter is that all basic particles (e.g., electrons) are identical to each other. If we have a system of these identical particles, such as the electrons in an atom, the system cannot be made different by a permutation of the particles. (By analogy, you might think of permuting three different colored balls to obtain six color patterns, as opposed to permuting three green balls and obtaining the same color pattern every time.) As you can see, this rule imposes a new kind of symmetry on the system, a much higher symmetry. One result of this high symmetry is that no two electrons of the system can be in the same state. Every electron in the system must differ from the rest in some way (a different energy, a different angular momentum, etc.). This result is known as the Pauli exclusion principle. The exclusion principle is responsible for all the different electron configurations of all the different elements. Without the exclusion principle, every element would have electrons that look like the electron in the hydrogen atom, and they would all behave chemically just like hydrogen. The wonderfully diverse chemical behaviors of the elements, and the order in these chemical behaviors seen in the periodic table (see chapter 2), are all due to the exclusion principle. The order imposed on chemistry by the periodic table is the result of an underlying symmetry.

Our last example comes from mathematics, and appropriately involves the invention of the group concept. This example deals with algebraic equations, such as the quadratic equation. The highest power of the variable (call it x) in a quadratic equation is the square of x. But we might have equations with the cube of x, or the fourth power of x, or any power of x we wish. This kind of equation is called a polynomial equation. In 1832, the 21-year-old Evariste Galois was investigating the properties of polynomial equations in a very general way. He discovered that the permutations of the solutions of the equations formed a group. This group preserved the algebraic relationships of the equations, much as we have seen groups of transformations which preserve the appearance of geometric figures. But symmetry groups for geometric transformations didn't yet exist; Galois coined the term "group" to describe the structures he had found. Galois had discovered a symmetry deeply hidden in the structure of algebra, from which many valuable results could be deduced (he was able to prove, for example, that no general solution can be written down for equations of the fifth power or higher). Galois summarized his results in a long letter that he desperately wrote during the night before he went to his death in a duel. Generations of mathematicians and scientists have worked to unravel the profound ramifications of his discovery.

FOR FURTHER READING

Symmetry, by Hermann Weyl, Princeton University Press, 1952.
Crystals and Crystal Growing, by Alan Holden and Phylis Morrison, MIT Press, 1982 (reprint; originally Doubleday, 1960).
Symmetry Discovered, by Joe Rosen, Cambridge University Press, 1975.
A Symmetry Primer for Scientists, by Joe Rosen, John Wiley and Sons, 1983.
Fearful Symmetry, by Ian Stewart and Martin Golubitsky, Blackwell Publishers, 1992.
Icons and Symmetry, by Simon Altmann, Oxford University Press, 1992.
Website: http://forum.swarthmore.edu/geometry/rugs/

EPIGRAPH REFERENCES: C. F. von Weizsacker, *The World View of Physics*, University of Chicago Press, 1952, p. 21. Hermann Weyl, *Symmetry*, Princeton University Press, 1952, p. 145.

Chapter 19

THE STRAIGHT AND NARROW: LINEAR
DEPENDENCE IN THE SCIENCES

> Moreover the simplest derived geometrical concepts,
> to which here belong especially the line and the plane,
> correspond to those which suggest themselves most
> naturally from the logical standpoint.
> *(Hermann Weyl)*

ABASIC QUESTION in almost any science is this: how does one thing depend on another? In physics, for example, we might ask how an object's position depends on time, or how a current depends on voltage. In chemistry, we might ask how the rate of a reaction depends on temperature, or how the reactivity of a metal with an acid depends on the pH of the acid. And in biology, we might ask how the metabolism of an animal depends on the amount of some hormone in its blood, or how the growth of a plant depends on the amount of rainfall it receives. These few examples could be multiplied almost without end. Sometimes, how one thing depends on another is quite complicated, but in this chapter we are interested in a very simple kind of dependence. Mathematically, we say that one variable depends on another variable, or that one variable is a function of another variable. A mathematical function is just a way of specifying the dependence. Again, the functional dependence may be simple or it may be complicated, and in this chapter we'll look at the simplest case possible: the linear function, otherwise known as a straight line.

§1. BASIC IDEAS

To understand what linear dependence means, let's start with an example that is familiar to most people, namely, working for an hourly wage. If you are working at a job that pays $12.00 per hour, then you'll make $12.00 working for one hour, $24.00 working for two hours, $36.00 working for three hours, and so on. We can summarize all of the particular cases in a single statement simply by saying that the amount of money you make is equal to the number of hours you work multiplied by $12.00 per hour. This can be written in the form of an equation, money = ($12.00 per hour)(hours).

The amount of money you make is said to be directly proportional to the amount of time you work. "Directly proportional" simply means that one quantity is equal to another quantity times some constant number (called the proportionality constant). In our example, the proportionality constant is $12.00 per hour. Of course, the proportionality constant might be different; for example, you might get a raise to $14.00 per hour. The amount of money you make is still linearly dependent on the number of hours you work, but you make more money now (after your raise) for the same amount of time worked. The proportionality constant, in this case, is a measure of how rapidly you accumulate money. We now have two different ways to say the same thing. A variable that depends linearly on another variable is, by definition, directly proportional to that variable; these two relationships are identical. What meaning does the word "linear" have in this context, and why do we use it as a synonym for direct proportionality? The word "linear" comes from the word "line." To see why that's appropriate, look at the graph in Figure 27. The amount of money you make is plotted on the vertical axis versus the number of hours you work plotted on the horizontal axis. As you see in Figure 27, the graph of a direct proportionality relationship is in fact a straight line. So, we call the relationship linear.

The solid line in Figure 27 is a graph of the money you make before your raise, while the dotted line is after your raise. They are both straight lines, but the line for $14.00 per hour has a steeper angle with the horizontal. The angle that a straight line makes with the horizontal axis always depends on the proportionality constant of the graphed relationship. The larger rate of pay having the steeper angle is no accident. For this reason, the proportionality constant is called the slope of the graph (e.g., the slope of the steeper line is $14.00 per hour).

All of these ideas are quite general. Instead of the money you make, think of some arbitrary quantity y; instead of the hours you work, think of another arbitrary quantity x; and instead of your wages, think of any arbitrary constant m. Instead of the previous equation, we now have

$$y = mx.$$

This may look more abstract, but $y = mx$ really isn't any more complicated than our simple example. The advantage of this more abstract version is that these symbols can now stand for anything we want. Let's illustrate the point with one more simple real-life example of a linear relationship. If you are traveling on the highway in a car that gets a gas mileage of 42 miles per gallon, the distance you travel is directly proportional to the

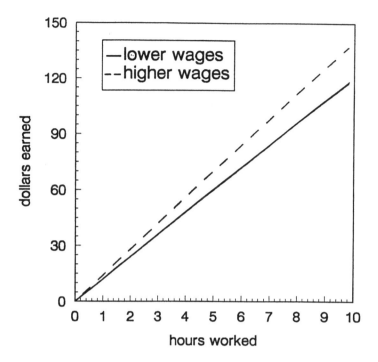

Figure 27. Graph of money earned at $12.00/hour (solid line) and $14.00/hour (dashed line), plotted against number of hours worked. The dependence of money on time worked is linear; the proportionality constant is the slope of the line.

amount of gasoline you use. In this example, y is equal to the distance you go, x is equal to the number of gallons of gasoline used, and $m = 42$ miles per gallon.

We can add a constant to our linear equation and still have a straight line with the same slope. Instead of just the money you make, for example, you might be more interested in your total savings. In that case, you add the (linearly increasing) money you are earning onto the amount of money you had to start with. Graphically, the effect of adding a constant is to shift the entire straight line vertically upward or downward on the graph. Our modified linear equation might look like

$$y = mx + b,$$

where b stands for the value of the constant we're adding. This constant is sometimes called the y-intercept of the graph (because $y=b$ when $x=0$, which is the y-axis).

Regardless of the particular value of m, our main interest may simply be that y is in fact directly proportional to x. In other words, y varies linearly with x, and this linear dependence is often the crucial information concerning some scientific phenomenon; the details of the proportionality constant may be less important to us. Scientists emphasize the importance of the functional dependence itself by using the symbol "$\propto$" which means "is proportional to." So, instead of using an equation, we can simply write

$$y \propto x$$

to express the fundamental idea we're interested in. To actually find numbers for y that correspond to numbers for x, however, we obviously need the proportionality constant.

§2. Examples of Linear Variation in the Sciences

Our primary motivation for discussing linear dependence is its usefulness and widespread application in the sciences. The property that makes linearity so useful is that it's the simplest functional dependence that two variables can have. (Even if there is *no* dependence, meaning one variable remains constant while the other changes, this too is a straight line, with a slope of zero.) A process or phenomenon that is governed by a linear relationship is easy to analyze and to understand. Fortunately, many such linear relationships are found in nature.

Constant Velocity

A simple example is motion with a constant velocity. Constant velocity implies a steady speed and direction, moving the same distance in each equal time interval. (If you move 25 meters during each second, for example, you have a constant velocity of 25 m/s; this is about 56 miles/hour.) The distance you travel is directly proportional to the time you've been traveling. The constant of proportionality in this case is the velocity. So,

$$\text{distance} = (25 \text{ m/s})(\text{time})$$

or, more generally,

$$x = vt,$$

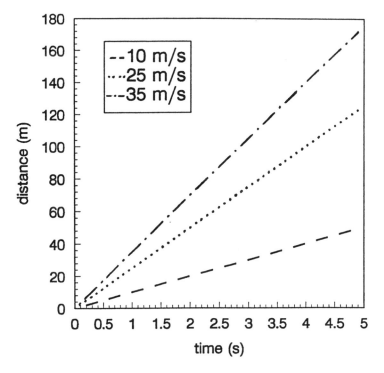

Figure 28. Graphs of constant velocity motion. The slope of the line is equal to the velocity in each case.

where x stands for the distance, t stands for the time, and v stands for the velocity. Graphs of distance versus time are given in Figure 28 for velocities of 25 m/s, 10 m/s, and 35 m/s, in order to illustrate once again the relationship between the appearance of the line and the numerical value of the slope (i.e., velocity).

In both our first example (involving money) and our last example (involving distance), the relevant quantity changes linearly with time. Put differently, both cases are concerned with a constant rate of change for some quantity. A constant rate of change is a fairly common application of the idea of linear variation in the sciences, and also in more general real-life situations. Another example might be a steady rainfall, in which the water level in a rain gauge increases linearly with time. But not all linear variations have to do with time rates of change, as our next examples show.

Density

Consider the relationship between mass, volume, and density. By definition, the density of a substance is its mass per unit volume. In other words, the density of an object is the mass of this object divided by its volume. This relationship can be written as

$$\rho = m/V,$$

where ρ is the density, m is the mass, and V is the volume. (ρ is the Greek letter rho.) Multiplying both sides of the equation by V, we have the equivalent form

$$m = \rho V.$$

For any particular substance (characterized by some density), the mass varies linearly with the volume. In a graph of mass versus volume, the slope of the straight line would be equal to the density of the substance. We see that the mass is directly proportional to the volume of an object, while the density is a property of the substance that the object is made of (and is unaffected by the size or shape). Of course, for a set of objects having the same volume, but made of different materials, the mass of each object is directly proportional to its density. Incidentally, it's interesting to note that, based on the material discussed in chapter 16, the mass does *not* vary linearly with the characteristic size or with the surface area of an object.

Ideal Gas Law

Let's now look at the ideal gas law from chemistry. As an equation, the ideal gas law can be written as

$$PV = nRT,$$

where P is the pressure a gas exerts, V is the volume of the container the gas is in, n represents the amount of gas, T is the temperature of the gas, and R is a constant (known as the universal gas constant). Suppose we keep the volume of the container and the amount of gas fixed. The pressure of the gas is then directly proportional to its temperature. P varies linearly with T; the proportionality constant in this case is nR/V. A variety of linear dependences are implied by the ideal gas law. If we keep the pressure constant, for example, the volume varies linearly with the temperature for a fixed amount of gas. If both the pressure and the temperature are held constant, then the volume is directly proportional to the amount of gas (slope = RT/P). A $V \propto n$ dependence is, of course, quite sensible if you think about it. But all of these simple linear relationships are predicated on the rest of the variables in the equation being held constant. If more than two quantities can vary at once, then we lose the simple linearity.

Hooke's Law

Another example, from physics, is known as Hooke's law. Hooke's law is usually introduced in connection with the force that a spring exerts when you stretch it (or compress it). The law states that the force exerted on an object is directly proportional to the distance through which the object is displaced. For a stretched spring, this means that the spring pulls back with a force proportional to the distance through which it is stretched. If you pull twice as far, the spring pulls back twice as hard, and so on. As an equation, Hooke's law is written as

$$F = -kx,$$

where F is the force, x is the distance stretched (or compressed), and k is the proportionality constant. The minus sign is there because the spring pulls in the direction opposite to the direction of the displacement (a force acting like this is called a restoring force). What meaning does the proportionality constant k have in this case? If k is large, then a small displacement results in a large force (and the opposite is also true). Some thought then reveals that the proportionality constant k in Hooke's law is a measure of how stiff the spring is. One reason why Hooke's law forces are so interesting is that many different physical systems are governed by forces that (at least approximately) have this form. A diatomic molecule, a guitar string, an atom in a solid, a pendulum, and a floating object that bobs up-and-down, are all examples of systems having restoring forces linearly proportional to displacements from equilibrium (equilibrium is defined as the position where the force is zero). Another reason Hooke's law forces are interesting is that such forces, being linear, are simple. Because the forces are so simple, we're able to analyze their effects and predict the motions they cause (these motions turn out to be periodic oscillations).

Other Examples

Next, let's look at an example from biology. The amount of oxygen consumed by an organism is directly proportional to the amount of energy it uses in metabolic activities. As a particular example of this, the stomach uses energy to secrete digestive acids. A graph of the oxygen uptake by the stomach versus the rate of stomach acid secretion is a straight line. Approximate linearity (see §3) is often useful in biology. The flow rate of blood through the circulatory system, for example, is approximately proportional to the pressure drop in the system (see chapter 6).

Finally, we'll consider one more example from chemistry. The boiling point and freezing point of a liquid changes if something is dissolved in the liquid. The boiling point gets higher and the freezing point gets lower. This lowering of the freezing point is familiar to everyone who has used salt to melt ice from sidewalks and roadways in the winter. The amount by which the temperature (of the freezing or boiling point) changes is directly proportional to the concentration of the solution. The constant of proportionality, in this case, depends on the identity of the components making up the solution.

§3. Approximate Linearity

As we see in these examples, many phenomena in the sciences are linear, which is one reason why linearity is important. Another reason is that certain phenomena, which are not really linear, can still be considered approximately linear (at least for some range of the variables). In other words, they are almost linear. (Approximate models are discussed more thoroughly in chapter 6.) In such cases, we can exploit the inherent simplicity of linear dependence in our analysis. As a simple example of approximate linearity, reconsider the gas mileage of your car. We said in section 1 that the distance you drive is linearly proportional to the amount of gasoline you use. But this isn't quite right. Some of your journey might be uphill, where your gas mileage is lower; highway driving gives you better mileage than city driving; and so on. The "constant" of proportionality actually varies a bit as driving conditions change. The relationship is only approximately linear. We've already seen a similar example from science, namely the relationship between blood flow rate and pressure drop in the circulatory system. Once again, the proportionality constant (resistance to blood flow) varies somewhat with the pressure drop, instead of being a genuine constant.

An interesting example of approximate linearity is shown in Figure 29, where atomic weights of elements are graphed versus their atomic numbers. The atomic number of an element is the number of protons the element has. The atomic weight depends on the number of neutrons and protons together (they have similar masses). The number of protons and the number of neutrons are roughly equal in each element, but not exactly equal. For this reason, the atomic weight is roughly proportional to the

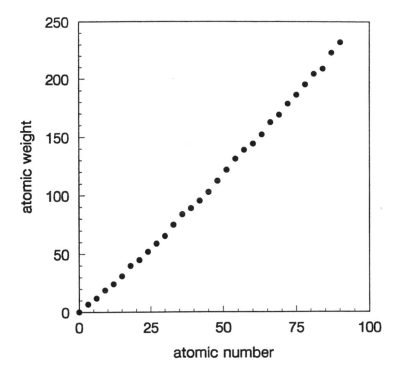

Figure 29. Graph of the relationship between atomic weight and atomic number in the elements, showing an example of approximate linearity. Along with the small amount of scatter, notice the upward curve at high atomic number (seen more easily with a straightedge or by looking at the plot along a glancing angle).

atomic number (with *m* approximately 2), but not exactly proportional. This relationship had interesting historical consequences. When the periodic table was first worked out (see chapter 2), the periodicity was in the atomic weights (which were known experimentally) rather than in the atomic numbers (which had not yet been invented as a concept; protons, neutrons, and nuclei would not be discovered for many years).

Our last example, from biomedical work, has important public policy ramifications. When an organism (e.g., a human) is exposed to some chemical (such as a medicine or a toxin), and the chemical has an effect, this effect is called a response to the chemical. The amount of the chemical to which the organism is exposed is called the dose. We can certainly expect some relationship between the dose and the response. If the response is quantified in some way, we can make a graph of the response versus the dose, sometimes called a dose-response curve. It's not uncommon for the dose-response curve to be approximately linear, espe-

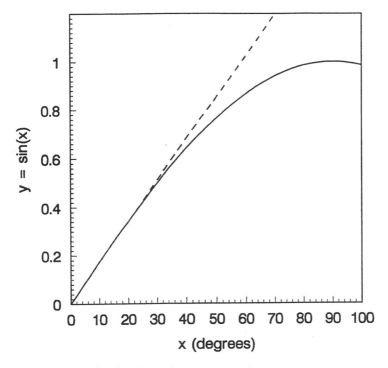

Figure 30. Graph of a clearly nonlinear relationship (a sine curve), along with a straight line that is an excellent approximation for part of the curve.

cially if the dose is not too high or too low. Questions about the dose range over which we can correctly assume the linear response approximation are important. For example, the toxicity of chemicals and radiation studied at relatively high doses is assumed to be linear down to low doses, where empirical studies are difficult. If the linear approximation breaks down at low doses, the danger of exposure might be underestimated or overestimated.

Many mathematical relationships that are not linear can also be considered approximately linear over some range. In Figure 30, we see a graph of a relationship that is decidedly nonlinear (it's actually a trigonometric relationship known as a sine curve). On the same graph, we plot a straight line. The straight line and the sine curve are virtually identical up to about 20 degrees. Most curves are nearly linear over some range of variables that is small enough, giving us an insight into why approximate linearity is common in nature. The opposite is also true: Most linear relationships break down if the range of variables is too extreme. The ideal gas law, for example, breaks down when the pressure is very high or the temperature is very low (see chapter 6; the density becomes high in both of these cases, and it's not surprising that the ideal gas law breaks down with the gas on the verge of becoming a liquid). Hooke's law also breaks down if the displacements become too large (which again is not too surprising, since you know that you can't stretch a spring indefinitely). These breakdowns do not detract from the usefulness or wide applicability of linear relationships. Any relationship in science is only valid within some proper domain of applicability.

Linear relationships are among the simplest relationships possible, and we've now seen that many phenomena in nature are linear (plus many more are at least approximately linear over some range). For these reasons, it's a common practice to assume that two variables are linearly related if we don't have any other information. This assumption can be quite useful when you try to make an order-of-magnitude estimate of some unknown quantity (see chapter 8). On the other hand, caution is advisable when drawing conclusions based on assumed linear dependence, as we saw in the case of dose-response relationships. Some phenomena turn out to be very nonlinear, and assuming linearity can lead to incorrect conclusions. Assumed linear dependence is an intelligent working hypothesis, but it needs to be checked by empirical tests. The simplicity of nature is remarkable, but not infinite.

EPIGRAPH REFERENCE: Hermann Weyl, *Philosophy of Mathematics and Natural Science*, Atheneum, 1963, p. 69.

Chapter 20

THE LIMITS OF THE POSSIBLE: EXPONENTIAL GROWTH AND DECAY

> The reader can suspect that the world's most important arithmetic is the arithmetic of the exponential function.
> *(Albert A. Bartlett)*

HOW THINGS CHANGE with time is a central question in the sciences. Some examples: how the shape of a riverbank changes with time; how a chemical reaction rate changes with time; how currents and voltages in an electrical circuit change with time; how a population of animals changes with time; how air temperature and rainfall change with time; how the inflation rate of the economy changes with time. These examples represent virtually all of the natural sciences (chemistry, biology, physics, geology, meteorology) and even include a social science (economics). There are many ways in which things can vary with time, some simple and some complicated. In chapter 19, we looked at linear changes with time, which are quite simple because the rate of change is constant. In this chapter, we'll look at a time variation that is somewhat more complicated, but exceedingly interesting and important. Instead of remaining constant, the *rate of change is proportional to the amount of the time-varying quantity*. We'll explore what this statement means and why it's so very important as we go along. Time variations of this sort are known as exponential changes. Exponential variation occurs for a variety of natural processes in many different sciences. We also find it in certain human social constructions, namely financial transactions. Such widespread occurrence is enough by itself to make exponential variation worthy of examination. As we'll see, however, the importance of exponential variation goes beyond academic interest; the future health of the human race may depend on how well we understand the implications of exponential growth.

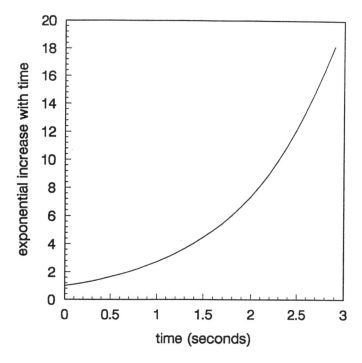

Figure 31. Graph of a quantity that grows exponentially with time.

§1. EXPONENTIAL DEPENDENCE

Exponential Growth

The term "exponential dependence" is really just a shorthand way to specify a particular kind of change. Although there are fancy mathematical definitions we could use, the easiest way to understand exponentials is to look at a graph. Figure 31 is a graph of a quantity growing exponentially with time, and this graph warrants careful examination. What properties does the exponential curve have? The first property you might notice is that the quantity grows with time. But you also see that the quantity starts out growing very slowly, and grows at an ever-faster rate as time goes on. In fact, the rate of growth is proportional to the size of the quantity at any point in time. *The more the quantity grows, the faster it's rate of growth is.* This is one of the central properties of exponential growth. To really appreciate the implications of this behavior, let's graph the quantity in Figure 31 over a somewhat longer period of time. We've done so in Figure 32, with the amount of time doubled from 3 s to 6 s on the

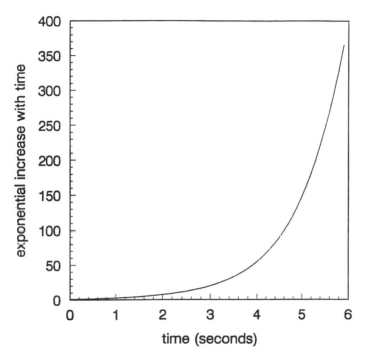

Figure 32. Graph of the same quantity shown in Figure 31, but over a longer time period; the portion of this curve for the first 3 seconds is identical to the entire curve in Figure 31.

horizontal axis. Look carefully at the numbers on the vertical axis, and you'll notice that the entire curve in Figure 31 is indeed the same as the first three seconds of the curve in Figure 32. What really draws your attention, however, is the huge increase that occurs after the first three seconds. The time has only doubled, but the exponentially growing quantity is about *twenty times larger*. Few people would have anticipated such a large increase. This difference between your expectation and the actual dramatic growth (which follows from the proportionality of growth rate to amount) is the main point to grasp here.

Exponential Decay

A quantity can shrink exponentially instead of growing. We usually refer to this process as exponential decay, and it's illustrated graphically in Figure 33. Once again, the rate of change is large when the quantity is large; but this time the quantity is shrinking, and so the rate of change

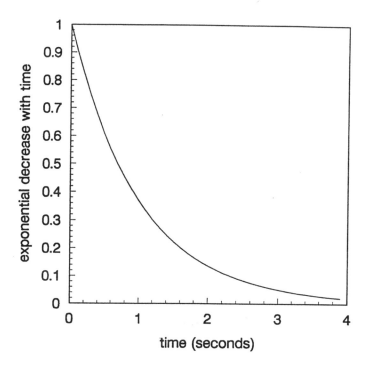

Figure 33. Graph of a quantity that decays (i.e., decreases) exponentially with time.

gets smaller as time goes on. After a long time, when not much of the quantity is left, the quantity shrinks very slowly. In §2, we'll look at several natural processes that behave this way.

Doubling Times and Half-Lives

An interesting property of exponential change is that the time needed to change by a factor of two remains constant for a given process. In other words, if some exponentially growing quantity doubled in size (e.g., from 25 to 50) after three hours, then it would also double in size (from 50 to 100) after another three hours. After yet another three hours (nine hours total), it would double again (from 100 to 200). And so on. Note carefully that this quantity grew as much in the last three hours as it grew in the first six hours. How long would it now take to grow as much as it grew in the entire nine hours? Just three more hours. This behavior is illustrated in Figure 34, which is based on the previous example. The characteristic time it takes for the quantity to grow by a factor of two is called the doubling time (the doubling time is three hours in our example and in

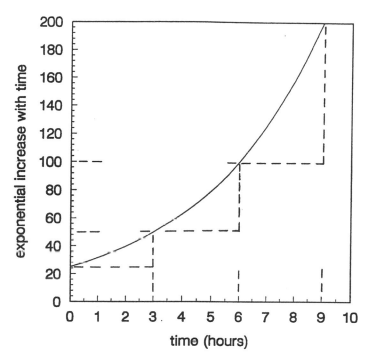

Figure 34. Graph of an exponentially growing quantity that has a doubling time of 3 hours.

Figure 34). It will always be the case for exponential growth that *a quantity increases as much in its final doubling time as it has previously increased in its entire history.* Exponential decay processes also have a characteristic time. In this case, of course, the quantity decreases by a factor of two instead of increasing. During each characteristic amount of time, the quantity shrinks to half of the size it was at the beginning of that time period. The name for the characteristic amount of time in this case is the half-life of the decay process. After one half-life, a quantity has decreased to half its original amount; after two half-lives, the quantity is one-quarter its original amount (half of a half); after three half-lives, the quantity is one-eighth its original amount (half of one-quarter); and so on. The actual amount of the decrease is less for each successive half-life because there is less to start with at the beginning of each successive time period. The meaning of a half-life is illustrated in Figure 33, where the half-life is about 0.7 second. Choose any 0.7 s interval on this curve, and you can verify for yourself that the quantity decreases to one-half of its value at the start of that interval.

§2. Some Examples

Population Growth

One of the classic examples of exponential growth in science is from biology: the growth of a population. Our reasoning applies to an organism that is not subject to constraints on available resources. For example, imagine a single-celled organism (such as a bacterium or an amoeba), which reproduces by cell division. Imagine also that the organism is sitting on a very large supply of nutrients and reproducing as fast as it can. At first, there is only one cell. This cell divides and then there are two cells; these two cells divide and then there are four cells; the four cells divide and there are eight cells; and so on. The number of cells (i.e., the population of the organism) doubles after every cell division. As we have seen, growth that is characterized by such a doubling of the population in given time periods is exponential growth. You may object that this scenario is unrealistic. Eventually, the nutrients will become depleted, or the waste products will slow the growth rate. These objections are absolutely correct. The growth will be exponential only while there are no such constraints, a situation that must surely be temporary. Indeed, one of the major points of this chapter is that exponential growth can't possibly be sustained for long periods of time. As long as these kinds of constraints are absent, however, the population growth can and will be exponential growth. Although the population doubling is easiest to see for single-celled organisms, which grow by cell division, exponential population growth is in fact typical of many creatures, including humans. Any population that increases by a certain *percentage* each year (as opposed to a certain fixed number each year) is an exponentially growing population. For example, the human population on earth has recently been growing at a rate of roughly 2 percent per year. If this growth rate stays the same, the population will increase exponentially with a doubling time of about 35 years. In less than 250 years, the world population would be more than *750 billion* people!

Cooling

A simple example of an exponentially *decreasing* quantity is the temperature of an object that cools by losing heat to its surroundings. Obviously, the object starts out at its maximum temperature, which is also the maximum temperature difference between the object and its environment. Your experience tells you that hotter things cool at a faster rate. In fact, the rate of cooling turns out to be directly proportional to this temperature difference between the hotter object and its cooler surroundings. As the object cools, and the temperature difference decreases, the rate of cooling slows down (a process that you've probably noticed if you've ever waited for a cake to cool down so you could eat it). As we've seen, this property (namely, a rate of change that is proportional to a changing quantity) is exactly the property that characterizes exponential change. So, the temperature difference between a hot object and its environment decreases exponentially as the object cools.

Radioactive Decay

Another famous example of an exponential decrease is the decay of radioactive nuclei. The nucleus of an atom is called radioactive if this nucleus is unstable; in other words, if the nucleus can change into a different and more stable form (often a different element), instead of just sitting there already in a stable form doing nothing. Most elements are stable and nonradioactive, but several elements with high atomic numbers are radioactive and naturally decay into more stable states, emitting radiation as they do so (which is the reason they are called radioactive). There are also unstable isotopes (versions of an element with a different number of neutrons) of many elements, and these also undergo radioactive decay. Now, the probability that a radioactive nucleus will decay is the same for every atom of a given element. The number of nuclei that decay in a given time is simply this probability times the number of nuclei present in the sample. The number that decay in a given time is the decay rate. Since the decay probability is just a constant, we see that the radioactive decay rate is proportional to the amount of radioactive material still left in the sample. Once again, this condition is the hallmark of exponential change. The remaining amount of a radioactive substance decreases exponentially with time.

You may have heard the term "half-life" used in connection with radioactivity in newspapers and magazines (radioactive substances feature prominently in a variety of public policy issues, such as nuclear waste disposal, medical applications, reactor leakages, and so on). We now see clearly what half-life means in this context: the half-life is the time it takes for one-half of the radioactive sample to decay into a stable species. Because the decay is exponential, the rate slows down as the sample shrinks. During the second half-life, only one-quarter of the original amount is lost, and so on. An interesting application of this principle is the technique of radiocarbon dating, used in archaeology to date primitive artifacts. A radioactive isotope of carbon (carbon-14) decays into stable nitrogen with a half-life of about 5700 years. By measuring the amount of carbon-14 remaining in a piece of wood or bone, for example, we can determine the number of years that have passed since the tree or animal died (and thereby stopped replenishing its carbon). Because we know the form of the exponential decay curve, the amount of remaining carbon-14 determines the number of years the sample has been decaying. In this way, the age of the object can be ascertained fairly accurately.

Compound Interest

Our last example of exponential growth is not taken from the sciences. Instead, this example is financial. Even though the phenomenon is not scientific, a discussion of compound interest is worthwhile because it's a familiar and important part of our day-to-day lives. Suppose we invest money and get a 5 percent return on our investment. At the end of some time period (for simplicity, let's say a year), we have an amount of money equal to the original investment plus 5 percent of that investment. During the next year, we get a 5 percent return on this new and greater amount of money. At the end of the second year, we've made more money than we did during the first year; this new and greater amount is now added to the total. During the third year, we get our 5 percent return based on this new (larger) total. And so on. Since the total increases *by a greater amount each year*, and since this ever-increasing total serves in turn as the basis for the next year's increase, then the rate of growth is increasing with the worth of the investment itself. Once again, we see the central characteristic of exponential growth in this process. If you work out the mathematics of compound interest, you find that the growth is indeed approximately exponential. This implies that your investment will grow only slowly for the early years, but will grow at quite a substantial rate after many years have passed (assuming you don't spend any of the money). All of this reasoning, incidentally, applies equally well to several other financial cases. Economic inflation, for example, increases the amount of money needed to purchase goods in the same way that compound interest increases the amount of an investment. If salary increases are awarded as a percentage of current salary, then the salary also increases in this manner. And growth in the economy itself (e.g., the gross national product) will be exponential if such growth is a consistent percentage of the current value, which is what economists and politicians typically strive for.

§3. Societal Implications

Exponential variation has fascinating mathematical characteristics and is certainly worth understanding for these alone. In addition, we've seen that a wide variety of natural phenomena all obey an exponential dependence law of some sort. Actually, there are many more examples than we've looked at here; transistor currents, chemical equilibria, capacitor charging, and a host of others could have been included. In this sense, exponential variation constitutes an underlying commonality across different scientific disciplines, which is really our main interest here. But there is another reason for understanding exponential variation. A variety of issues at the science/society interface turn on the concept of exponential change. Before leaving the subject, let's explore some of these issues in more detail.

We've already mentioned the problem of nuclear radioactive waste disposal. An understanding of the concept of a half-life, and what this means in the context of an exponential decay law, is necessary in order to understand the nature of this problem. A clear understanding of the problem is needed before we can assess the proposed solutions to the problem. Of course, many other facets of the problem need to be understood and accounted for (effects of radiation on human health, geologic stability of proposed sites, etc.). But to even begin thinking about these various facets of the problem would be difficult without a basic understanding of exponential radioactive decay.

Another problem, which we've also mentioned briefly already, is human population growth. In order to have a decent life, humans need resources such as food, space, energy, and so on. If there are too many people on earth, it is simply not possible to provide these resources to everyone. How many people are too many? We don't yet know in detail the answer to this question, but the major point to make here is that the details of the answer are almost irrelevant if population growth continues to be exponential. Because the rate of growth increases with the size of the population itself, a disaster is obviously inevitable; the details only tell us whether this disaster will be sooner or later. Now, when I say that a disaster (for example, mass starvation) is inevitable, I mean that it's inevitable *if* the growth continues to be exponential. Hopefully, humanity will be able to curb this exponential population growth by volition before the growth is stopped in a very unpleasant manner by forces we cannot control. A closely related problem is exponential growth in the use of natural resources, such as metal ores, fossil fuels, and so on. Such growth in the use of resources (which is historically well documented) is partly due to population growth, and partly due to economic growth (these are related but not identical). The basic problem, of course, is that the total amounts of the resources are fixed and finite, while the rates at which the resources are consumed increase without end. The problem is especially crucial for fossil fuels, which are nonrenewable and cannot be recycled. The obvious conclusion is that someday these resources will be completely depleted and thereafter gone forever; the only question is when this will happen. In thinking about this issue, the importance of understanding exponential growth can't be stated too strongly. For example, someone might argue that we have greatly underestimated the available oil reserves on earth and that there is ten times as much oil as we had previously thought. This statement ("ten times as much") sounds very impressive. However, the seemingly large increase in available resources would be used up in little more than three additional doubling times. For a growth rate of just a few percent per year, this is about one human lifetime. What if we discovered instead that we had one hundred times as much oil as we had thought? This seemingly vast supply of oil would all be used up in less than seven additional doubling times, that is, about two human lifetimes.

For Further Reading

"Forgotten Fundamentals of the Energy Crisis," by Albert A. Bartlett, *American Journal of Physics*, vol. 46, p. 876, September 1978.

e: The Story of a Number, by Eli Maor, Princeton University Press, 1993.

"Exponential Growth and Doubling Time," in *Conceptual Physics*, by P. G. Hewitt, HarperCollins, 1993, p. 705.

Epigraph reference: Albert A. Bartlett, in *American Journal of Physics*, vol. 46, 1978, p. 877.

Chapter 21

IN THE LOOP: FEEDBACK, HOMEOSTASIS,
AND CYBERNETICS

> The feedback of voluntary activity is of this nature. We
> do not will the motions of certain muscles, and indeed we
> generally do not know which muscles are to be moved to
> accomplish a given task. . . . Our motion is regulated by
> some measure of the amount by which it has not yet
> been accomplished.
> *(Norbert Wiener)*

THE PRESENCE of feedback, in both natural systems and technological systems, is often at the heart of how these systems function.
What do we mean here by the word "feedback?" Before attempting
a formal definition, let's consider a simple example. If we turn a space
heater on in a small room, the temperature in the room will simply continue to rise until the room becomes quite warm. There is no feedback in
this case. But now suppose that the heater contains a temperature sensing
device, which lowers the heat output if the temperature rises (and increases the heat output if the temperature goes down). In other words,
suppose we have a thermostat. We see very different behavior this time;
the temperature will rise to some moderate value, and subsequently
fluctuate about this value. We say that there's a feedback operating in
this case, because the effect of the heater (the temperature rise) has an
influence on its cause (the heat output). Part of the effect has been fed
back to the cause, and modified it. We can now see what feedback means
in general: Some input (or cause) gives rise to some output (or effect),
but a part of the output is fed back to the input so as to influence the
behavior of the system. The part of the output that goes back to the input
is termed feedback, and the whole system (including the input, the output,
and the feedback) is often referred to as a feedback loop. Figure 35 shows
a schematic picture of a feedback loop (you can see there why it's called
a loop). Similar terminology is sometimes used in everyday conversations.
If we request feedback on a piece of writing, for example, we want a
response to an action, which can modify further action (in this example,
a second draft).

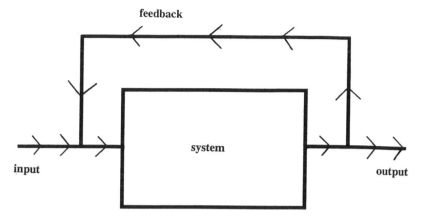

Figure 35. A schematic illustration of a feedback loop. The input to the system gives rise to some output. Part of the output feeds back to influence the input, modifying the behavior of the system.

§1. POSITIVE FEEDBACK AND NEGATIVE FEEDBACK

The mechanism of a feedback loop may be simple or complicated, depending on the details of the system. Regardless of the details of the mechanism, though, all feedbacks can be classed into two broad categories: positive feedback or negative feedback. The meanings of these terms in science are somewhat different from their everyday conversational meanings. Positive feedback means that the output reinforces the input that caused it. Conversely, negative feedback tends to decrease the input. The thermostat example, with which we started the chapter, is an example of negative feedback, whereas a heater that *increased* its heat output as the temperature rose would illustrate positive feedback. What would happen if we had such a positive feedback heating system? As you can see, positive feedback is not always desirable. In fact, positive feedback often leads to an uncontrolled (and therefore undesirable) result. The classic example of unwanted positive feedback is putting a microphone in front of a speaker. Any small random sound picked up by the microphone is amplified by the audio amplifier and fed to the speaker, where it's picked up by the microphone again and sent (now louder) to the amplifier. The amplifier again feeds the sound to the speaker (even louder) and hence to the microphone, and so on. This process continues until we hear that extremely loud and obnoxious wailing sound that everyone is familiar with. Positive feedback can also cause a decrease instead of an increase, as long as the decrease is magnified by the feedback loop. Roughly speaking, this happened during the stock market crash of 1929. When falling stock prices frightened investors, they sold their stocks, causing another fall in stock prices. Even more frightened, investors sold yet more stocks, resulting in even lower prices, and so on.

Negative Feedback and Stability

Negative feedback, on the other hand, usually leads to stability. Because it provides stability, negative feedback is often used in engineering control systems. A simple and familiar example of a control system is the thermostat, described at the beginning of the chapter. A rudimentary thermostat (like that found in most homes) simply turns heat on when the temperature goes below a certain level, and turns it off again when the temperature is high enough. Another example is the cruise control on some automobiles, which uses feedback to maintain a steady speed by increasing and decreasing the amount of gasoline, depending on whether the car is slowing down or speeding up. In the nineteenth century, steam engines had a control device called a governor to perform this task, using an ingenious mechanical feedback method.

Electrical engineers have become highly skilled in using negative feedback to make electronic circuits do what they want. An amplifier is a circuit that takes some input voltage and produces an output voltage that is just a magnified version of the input. For example, an audio amplifier takes the tiny voltage signals provided by a tape deck or CD player and produces a powerful output voltage that can drive the speakers. But if drifts and fluctuations occur in the output, we'll get a distorted and unpleasant musical sound. By making a fraction of the output negative and feeding this back to the input, we cancel out such drifts and fluctuations. In this way, we can make a highly stable and linear amplifier. Such amplifiers have a variety of important uses in sound reproduction, automated control systems, and high-precision scientific instruments.

Natural systems can also have negative feedback, which maintains their stability. In a biological system consisting of a predator and prey, for example, an increase in the prey population would result in an increase in the number of predators, since they would have more food to eat. But an increase in the number of predators would in turn reduce the growth of the prey population, because the prey would be eaten faster. Of course, the decrease in prey would then cause a decrease in the number of predators, because there would now be less food to eat. And so it goes, the net effect being relatively stable populations of both predator and prey. In a real ecosystem, of course, there will be a highly complex network of such feedbacks. On a global scale, the amount of oxygen and carbon dioxide in the atmosphere is partially regulated by feedback from the animals and plants that consume and generate these gases. If the amount of carbon dioxide (which the plants need to live) decreases, for example, the plants will grow more slowly and use less carbon dioxide, thereby slowing the decrease. One can only imagine the complexity of all these feedback processes in the entire global ecosystem.

Positive Feedback and Oscillations

Positive feedback also plays an important role in both nature and technology. We've seen how positive feedback leads to uncontrolled growth of the output, but in reality such unlimited growth can't last forever. Instead, one of two things might happen: either the system saturates, or else it lapses into oscillations. As an example of the former possibility, consider our insane thermostat, which called for more heat as the temperature rose. The heater would soon be producing all the heat it could, a situation known technically as saturation, and no further change would take place. The speaker/microphone case is also an example of saturation, since the noise is as loud as it can get (limited by the power of the amplifier). But as I said, the system may not saturate; instead, it may oscillate.

Oscillations are simply vibrations, or periodic to-and-fro motions. A simple example of oscillatory motion is a child swinging on a swing. Oscillations can be rapid or slow, and the measure of how fast the oscillations are is known as their frequency. In order to have an oscillatory output, a feedback system must have something special about its feedback loop: Instead of feeding back everything in the output, the loop only feeds back a part of the output that has some particular frequency. That frequency will be amplified to saturation, while nothing else is amplified at all. The resulting output consists of oscillations at that frequency. To visualize this process, think about the child on the swing. If the swing just has small random movements (which are a mixture of many different frequencies), then not much happens. Assume for the moment that we have some way to amplify one frequency and feed it back to the input, that is, push the swing with that one particular frequency. The swing is now going at *that* frequency, and if we amplify it again and again, the swing will soon be going high, starting from just small random motions.

It's not hard to design electronic feedback loop circuitry with this property (only feeding back a selected frequency), and such circuits do indeed have oscillatory voltage outputs. These circuit devices, called oscillators, are used in many important applications. Electronic music synthesizers, for example, use oscillator circuits to create sounds. The timing circuitry in computers and digital watches is another application of oscillators. In fact, the quartz crystals in so-called quartz watches are part of a feedback loop, and these crystals vibrate with a highly precise frequency. The watch uses this frequency to keep time. A similar mechanism occurs in the sound production of flutes and organ pipes. Air blowing across the mouthpiece of the flute or lip of the pipe enters into unstable turbulent vibrations (in other words, random movements up and down over the lip). The air column of the instrument, however, has a natural frequency with which it wants to vibrate (similar to a plucked string; the natural frequency depends on the length of the air column or string). A positive feedback loop couples the vibrating air column with the blown air, resulting in strong oscillations at the natural frequency of the tube and a pleasant musical tone.

§2. REGULATION AND CONTROL

Homeostasis

Individual organisms, notably the human body, regulate their functions using a host of negative feedback systems. Perhaps the most well-known example of this regulation is body temperature. We need to maintain body temperature within a fairly narrow range to stay alive, despite widely varying external temperatures and internal metabolic activity (yet another thermostat). The main mechanism for this control is the increase and decrease of perspiration, in response to increasing and decreasing body temperature. Shivering is used to produce extra heat if the temperature falls too low. Another excellent example of regulation in the body is the chemical composition of blood. Blood chemistry must also remain within narrow limits in order for us to stay alive. Some of the feedbacks are fairly simple, even if the actual physiological mechanisms are complex. The kidneys, for example, simply remove urea at a higher rate as the urea level in the blood increases. Not all of the feedback loops are so simple, though. The regulation of the blood's pH level (i.e., the acidity of the blood) requires a complex and interrelated set of feedback loops involving carbon dioxide, carbonic acid, hemoglobin in the red blood cells, and the action of the lungs in exchanging gases between the blood and atmosphere. The name given to all these types of regulatory processes, and the stable state they produce, is "homeostasis." As we've seen, homeostasis is essential to the continuation of life. Other examples of homeostatic regulation include hormone levels, hydrogen ion concentration, blood pressure and heart rate, calcium metabolism, and various functions of the parasympathetic nervous system.

Cybernetics

Feedback loops are also involved in the control of voluntary behavior. If I reach out to pick up an object, my eyes sense the position of my hand relative to the object. This information is used to constantly modify how I continue to move my hand. In persons who have suffered damage to the part of the brain that performs this feedback, any attempt to pick up an object results in wild and uncontrolled swings. In fact, we need to use negative feedback mechanisms just to hold still, and damage to these systems results in tremors. Feedbacks also operate in situations such as steering a car. We're constantly correcting for drift from one side to another. The feedback mechanisms are more subtle in these cases; the brain receives information from the senses, processes this information, and in response sends commands to the muscles. Once again, though, negative feedback is used to bring about more stable control. Norbert Wiener studied these kinds of feedback processes extensively. Wiener was a very influential scientist and mathematician; with his colleagues, he combined the study of feedbacks with the emerging sciences of information theory, computer science, and neurophysiology. A new science developed from this work, which he termed "cybernetics." Cybernetics is the general study of control and communication in both humans and machines.

A simple example of a cybernetic device is our familiar thermostat, where a human and a machine work together to control temperature. Extremely complicated versions of such cybernetic control systems have been designed for diverse applications, such as automated manufacturing techniques. These control systems employ sensors to detect positions, pressures, temperatures, and so on. Computers are used to process all the information from the sensors and make decisions based on the information; servomechanisms then carry out the commands of the computer, such as moving robot arms or opening and closing valves. Humans monitor the entire process, telling the computer how to make its decisions and formulating the goals of the process. The founding of cybernetics was not only intellectually important, but also had practical consequences. Industrial automation was a direct result of the early cybernetic studies by Wiener and his coworkers. The consequences of automation have been profound. Automation has increased efficiency and saved humans from doing dangerous jobs, but it has also caused many jobs to disappear, taken over by machines. Intellectually, the development of cybernetics has led to a new view of the world, more rich and complex than the simple cause-and-effect models that preceded it. The introduction of feedback into a system means that the effect and the cause can no longer be completely disentangled. The behavior of such a system is far more diverse and inter-esting (and complicated) than a system without feedback. Cybernetics has made science more aware of the presence of feedback in many systems and provided the tools needed to study these systems.

§3. COMPLEX FEEDBACKS

The earth's climate is affected by feedback loops, both positive and nega-tive. For example, a rise in the average temperature of the air results in more water vapor in the air (the reason it's more humid in summer than in winter). Water vapor is a greenhouse gas, which means that it traps the sun's heat in the atmosphere. Increasing the water vapor in the air further raises the temperature in a positive feedback loop. More water vapor in the air, however, causes more cloud formation; the clouds decrease the amount of sunlight reaching the earth, so there is also a negative feedback at work. Many such complicated feedbacks, working together, determine the earth's global climate. Questions about these feedback loops are a major source of difficulty in predicting the effects of adding more green-house gases (from fossil fuel combustion and other industrial processes) into the atmosphere (see chapter 10).

Feedback loops are created by a variety of different mechanisms in dif-ferent kinds of systems. We've seen at least three different kinds of feed-back mechanism so far. The simplest kind is just a fraction of the output fed back to the input (which occurs, for example, in electronic linear am-plifiers). A more complicated mechanism exists when the output is de-tected and then affects the input in response. A thermostat, controlling a heater in response to temperature changes, is an example of this. Finally, in some cases the feedback is in the form of information alone; this is particularly likely if a human or a computer is part of the feedback loop. Feedback may be linear (i.e., directly proportional to the output) or it may be nonlinear. Because it's simpler, the linear case is easier to under-stand (see chapter 19), and we can usually predict the output in such cases. If the feedback is nonlinear, however, the output may change in a complicated and perhaps unforeseen fashion; nonlinear systems can ex-hibit a rich variety of interesting behaviors (see chapter 17). An extreme example is the human brain. The neurons of the brain are connected to-gether in an indescribably complex array of nonlinear feedback loops, resulting in our mental processing.

Around the middle of the twentieth century, a number of thinkers felt that the old disciplinary structures of science were becoming outmoded. They developed a discipline known as general systems theory, in which the concept of feedback played a prominent role. Some proponents of general systems theory thought they could unite all of the sciences into a single unified discourse. While this never really came to pass, it's certainly true that the concept of feedback is valuable in many different sciences and that it unifies many seemingly disparate phenomena. Remarkably, we can use the same basic concept to discuss electronic circuitry, biological ecosystems, engineering control systems, global climates, flutes, tremors, and homeostatic regulation in the human body—and thermostats.

FOR FURTHER READING

Organism and Environment, by J. S. Haldane, Yale University Press, 1917.

Cybernetics, by Norbert Wiener, M.I.T Press, 1948 (first ed.) and 1961 (second ed.).

The Unified System Concept of Nature, by S. T. Bornemisza, Vantage Press, 1955.

"Integrative Principles in Biology," by Robert M. Thornton, in *Integrative Principles of Modern Thought*, edited by Henry Margenau, Gordon and Breach, 1972.

Introduction to Physiology, by H. Davson and M. Segal, Grune and Statton, 1975.

Physics and the Sound of Music, by John Rigden, John Wiley and Sons, 1985.

EPIGRAPH REFERENCE: Norbert Wiener, *Cybernetics*, MIT Press, 1965 (paperback version of second edition), p. 97.

Epilogue

SO, WHAT IS SCIENCE?

> Each had felt one part out of many. Each had perceived it
> wrongly. No mind knew all: knowledge is not the companion
> of the blind. All imagined something, something incorrect.
> (*From* The Blind Ones and the Matter of the Elephant,
> *a Sufi tale as retold by Idries Shah*)

I WOULD LIKE to end with some pithy aphorism that sums up in a few words my understanding of what science is. But I don't believe that's possible. Every brief attempt I've seen to define science has failed to capture some crucial element of the total picture. Even a longer attempt (like this entire book) can't even begin to include everything that's worth saying about science. Instead, I'll end with a story. When I was visiting a group of children (my daughter's fourth grade class) to do some science activities with them, one of our projects was to find out which things conduct electricity and which don't. Hooking together a battery and light bulb with some wires, the children inserted various objects between two wires and observed whether the light bulb was glowing or not. If the bulb lights up, the object is a conductor. Most of the conductors turned out to be made of metal (paper clips, key chains, orthodontic braces, etc.). But then the children had a big surprise: pencil lead (graphite) is an electrical conductor, even though it's not shiny and metallic-looking like the rest of their collection. Although I wasn't able to explain this to the children, we do understand why graphite is a conductor despite the dissimilarities between graphite and the metals. The explanation is largely based on concepts we've seen in chapter 1 and chapter 18. Basically, the unusual crystalline structure of graphite (consisting of two-dimensional sheets) accounts for the special properties it has.

This anecdote illustrates many aspects of how science works: starting with ideas and concepts you know, observing the world, trying different things, creating a coherent context, seeing patterns, formulating hypotheses and predictions, finding the limits where your understanding fails, making new discoveries when the unexpected happens, and formulating a new and broader context within which to understand what you see. Many of these same themes appear sprinkled throughout this book. If there is any single statement that encompasses as many of these themes as possible, perhaps it is this: Science is the active and creative engagement of our minds with nature in an attempt to understand. Beyond this broad generality lies the enjoyment of exploring a variety of particular paths in science.

EPIGRAPH REFERENCE: Idries Shah, *Tales of the Dervishes*, E. P. Dutton, 1970, p. 25.

INDEX

accuracy, 114
acetylene, 15
acid rain, 118–19
ad hoc explanation, 179
ad hominem fallacy, 104
Adams, J. C., 50
Airy, G., 50
algebra, 272
Alhambra, 269
amino acids, 4
amplifier, 297
analogy, 98
Andes, 20
applied science, 134
approximate linearity, 281–84
Archimedes, 55
argon, 29, 42–43
argument: converse of, 99; inverse of, 99–
  100
Aristarchus, 53, 65
Aristotle, 44, 54, 90, 190
arrow of time, 195, 235–36
art, symmetry in, 268–69
asthenosphere, 34
atmosphere, 42
atmospheric pressure, 45–48
atomic bomb, 150
atomic number, 281–82
atomic weight, 27, 281–82
atoms, 69
autocatalysis, 238

Babylonians, 52
background, 179
Bacon, Francis, 133, 135, 190
Bacon, Roger, 37, 130
band gap, 17–18
band structure, 17–19
Bardeen, John, 37
barometer, 42, 44–48
Bartlett, Albert A., 285
beehive, 268
begging the question, 105
behaviorism, 152, 156
Belousov-Zhabotinski reaction, 238
benzene, 14
big bang, 271

bilateral symmetry, 253–54, 266–67
billiard balls, 69, 74
biogeography, 19–20
black holes, 40–41
Bloch, Felix, 17
blood flow, 75–76, 281
Boltzmann, Ludwig, 234
bonding energies, 112
Boolean network, 238–40
Brahe, Tycho, 58–59
Brevais lattice, 262
Bridgman, P. W., 35, 191
broken symmetry, 264–65
Bruno, Giordano, 127
buckminsterfullerene, 259
Burckhardt, Titus, 162
burden of proof, 186
Burtt, E. A., 145
Bush, Vannevar, 133, 136
butterfly effect, 244

Capra, F., 131
carbon, 14
cards, 181, 183, 234–35
Cassirer, Ernst, 3, 195
catalysts, 226
cathode ray tube, 12
cats, 90–92
causality, 96–97, 192–93
cells, size of, 225
cellular automata, 240
Chandrasekhar, S., 209–10
chaos theory, 241–51
characteristic length, 223
Chardin, Teilhard de, 131
chemical clock, 237–38
chlorophyll, 258
circle, 53–54, 222, 254, 256
Clausius, Rudolf, 233
closed system, 236
cold fusion, 175–80
communication, 137–38
compound interest, 292
conditional proposition, 92
conductors, 18, 303
conflicts, 81–83
conservation laws, 271